Nanoelectronics

The book covers the latest advancements in the field of nanoelectronics, interaction of light with matter, optical properties, surface plasmons, nanophotonic and plasmonic applications, nanomaterials engineering in nanoelectronics, fabrication techniques, and applications. It presents the structure properties and device applications of carbon nanotubes, graphene, and silicene. The book further includes topics on perovskites and applications, quantum dots for optoelectronic, sensing devices, computing applications, and advanced applications in diagnostic and therapeutic tools.

This book:

- Introduces topics on quantum phenomenon in nanoelectronics, transistors in modern computing, spintronic devices, flexible and wearable electronics, and case studies and includes topics on development of hybrid material of one-dimensional (1D) and 2D carbon in nanomedicine, computed tomography, ultrasound, and stapled peptides drug.
- Presents physical and electrical properties of nanoscale materials, applications of carbon nanotubes such as batteries, supercapacitors, biosensors, photodetectors, and field emission sources and topics on the surface plasmon resonance, nanolasers, nanoantennas, photonic crystals, and luminescent materials.
- Discusses quantum dots, quantum electronic devices, nanosensors, quantum computing, device applications of perovskite materials, spintronics, drug delivery, and gene therapy applications and the scaling laws in nanoelectronics, materials for nano-FET, high-speed computing, and advanced nanoscale FET architectures,
- Highlights recent developments in the field of nanoscale devices based on graphene and silicene such as biosensing and gas sensing, thermoelectric devices, and optoelectronic applications.
- Explains the basic quantum theory, pizoelectricity in zinc oxide quantum dots, and nanosensing, includes the topics on data-storage and quantum logic gates.

It is primarily written for senior undergraduates, graduate students, and academic researchers in the fields including physics, computers, nanoelectronics, electrical engineering, electronics and communications engineering, nanoscience, and nanotechnology.

Nanoelectronics

Advanced Concepts and Device Applications

Edited by
Khurshed Ahmad Shah

CRC Press
Taylor & Francis Group
Boca Raton London New York

CRC Press is an imprint of the
Taylor & Francis Group, an **informa** business

Designed cover image: Shutterstock

First edition published 2026
by CRC Press
2385 NW Executive Center Drive, Suite 320, Boca Raton, FL 33431

and by CRC Press
4 Park Square, Milton Park, Abingdon, Oxon, OX14 4RN

CRC Press is an imprint of Taylor & Francis Group, LLC

ISBN: 9781032844817 (hbk)
ISBN: 9781032844824 (pbk)
ISBN: 9781003512899 (ebk)

DOI: 10.1201/9781003512899

Typeset in Sabon
by Newgen Publishing UK

Contents

About the editor

Khurshed Ahmad Shah is currently working as Associate Professor, PG Department of Physics, Sri Pratap College, School of Sciences, Cluster University Srinagar, Srinagar, Jammu and Kashmir, India. He obtained his doctorate in Physics from Jamia Millia Islamia, Central University, New Delhi, India, and Master of Science and Master of Philosophy through University of Kashmir, Srinagar, India. With over 15 years of teaching experience, Dr. Shah has effectively integrated technology into his instruction, utilizing the internet and developing customized algorithms to enhance student understanding. As Head of the Physics Department for over eight years, he initiated student visits to national research institutions and organized modern laboratory facilities, including an IT lab and a darkroom lab. His efforts in science outreach extend to school and college students through his special lectures on nanoscience, hands-on nanomaterial preparation sessions, and scientific model projects, nurturing future leaders in science and technology.

Dr. Shah has successfully led seven sponsored research projects as Principal Investigator, mobilizing over Rs. 2.0 crore in research funding. His efforts have culminated in the establishment of two research laboratories poised to produce hundreds of future Ph.D scholars. He has published extensively in high-impact journals, supervised Ph.D candidates, and organized numerous conferences, workshops, and seminars, enriching both academia and the broader scientific community.

Dr. Shah's groundbreaking work in carbon material synthesis has led to the development of novel catalysts and carbon nanotubes with enhanced properties suited for device applications. Notably, he introduced an innovative formula for determining the nature of carbon nanotubes, a formidable challenge in this field, and reviewed crucial elements of carbon nanotube synthesis, an article that has garnered over 700 citations. His advancements in modeling and simulation have led to the design of nanoscale sensors, heterostructures, and photodetectors based on one-dimensional (1D) and 2D materials, with promising applications in the electronics industry. Furthermore, in collaboration with IIT Ropar, Dr. Shah secured an Indian patent for a single-process method to produce four carbon nanostructures from walnut shell, including carbon quantum dots and graphene derivatives.

Dr. Shah co-authored six books and four book chapters in reputed international publications. In addition, Dr. Shah is a lead editor of book titled *Nanoscale Electronic Device Applications of Carbon Nanotubes, Graphene, Silicene, and Molybedinum Disulphide*, to be published soon by CRC Press. Besides, Dr. Shah is an editorial board member and reviewer of many scientific journals of leading publishing houses. Dr. Shah's involvement with the National Service Scheme (NSS) has had a profound impact, inspiring students to engage in community services like sanitation drives, wildlife protection, biomass management, and COVID-19 awareness. He has also actively promoted scientific temperament in Jammu and Kashmir (J&K), coordinating a wide array of science and technology events, from essay competitions to interactive sessions with prominent scientists, reaching over thousands of participants and earning him an appreciation letter from the J&K Science Technology and Innovation Council.

Dr. Shah's academic affiliations further underscore his professional standing. He became a Fellow of the prestigious Institute of Physics, London, in 2023 and holds memberships in esteemed organizations such as the American Physical Society and the International Association of Advanced Materials, Sweden. He was also inducted as a member of NASI in 2021. Further he has worked as Secretary, NPTG and QEP groups of Institute of Physics, London, UK. Among his numerous accolades are the Visiting Scientist Fellowships from Howard University and the Indian National Science Academy, as well as four fellowships from JNCASR, Bangalore, where he had the privilege of collaborating with Bharat Ratna Prof. C.N.R. Rao. Additionally, Dr. Shah's awards include the J&K State Young Scientist Fellowship, State Innovative Science Teacher Award, and the Jawaharlal Memorial Fellowship for Doctoral Studies, where he ranked first nationally.

Contributors

Chase Adams
Laser Spectroscopy Laboratory, Department of Physics & Astronomy, Howard University, Washington, DC, USA

Roweena Agnes
Department of Physics, SASTRA University, Thanjavur, Tamil Nadu, India

Saumika Bandopadhyay
Laser Spectroscopy Laboratory, Department of Physics & Astronomy, Howard University, Washington, DC, USA

Basanta Baral
Laser Spectroscopy Laboratory, Department of Physics & Astronomy, Howard University, Washington, DC, USA

Wisdom Benson
Laser Spectroscopy Laboratory, Department of Physics & Astronomy, Howard University, Washington, DC, USA

Sheenu Bhadauria
Department of Chemistry, Kasturbagram Rural Institute, Indore, MP, India
IET, Applied Science Department, Devi Ahilya Vishwavidyalaya, Indore, Madhya Pradesh, India

Lal Kumar Chandel
Department of Chemical Sciences, Maharaja Ranjit Singh College of Professional Sciences, Indore, Madhya Pradesh, India

Nandu B. Chaure
Department of Physics, Savitribai Phule Pune University (Formerly University of Pune), Pune, Madhya Pradesh, India

Sajad Ul S. Dar
PG Department of Physics, Sri Pratap College, School of Sciences, Cluster University Srinagar, Srinagar, Jammu and Kashmir, India

Asmita Deshpande
Shri Ramdeobaba College of Engineering & Management, Nagpur, Maharashtra, India

Peerzada F. Faroz
PG Department of Physics, Sri Pratap College, School of Sciences, Cluster University Srinagar, Srinagar, Jammu and Kashmir, India

S. Gomathi
Department of Chemistry, Periyar Maniammai Institute of Science & Technology, Thanjavur, Tamil Nadu, India

Brijesh Kumbhani
Department of Electrical Engineering, Indian Institute of technology, Rupnagar, Punjab, India

Priyanka U. Londhe
Department of Physics, Savitribai Phule Pune University (Formerly University of Pune), Pune, Maharashtra, India

Francis Mensah
Department of Natural Sciences, Virginia Union University, Richmond, VA, USA

M. Farooq Mir
Sri Pratap College, School of Sciences, Cluster University Srinagar, Jammu and Kashmir, India

Ulfat T. Mir
PG Department of Physics, Sri Pratap College, School of Sciences, Cluster University Srinagar, Srinagar, Jammu and Kashmir, India

Prabhakar Misra
Laser Spectroscopy Laboratory, Department of Physics & Astronomy, Howard University, Washington, DC, USA

Hariharan Rajalakshmi Mohanraj
Computational Materials Science and Nanodevices Simulation Laboratory, Department of Physics and Nanotechnology, Faculty of Engineering and

Technology, SRM Institute of Science and Technology, Kattankulathur, Tamil Nadu, India

H.P. Nagaswarupa
Department of Studies in Chemistry, Davangere University, Shivagangothri, Davangere, Karnataka, India

Ramachandra Naik
Department of Physics, New Horizon College of Engineering, Bangalore, Karnataka, India

M. Nidhishree
Department of Chemistry, Periyar Maniammai Institute of Science & Technology, Thanjavur, Tamil Nadu, India.

J.S. Nirmalram
Research & Development Cell, Department of Chemistry, PRIST Deemed University, Thanjavur, Tamil Nadu, India

Akash Ramasamy
Computational Materials Science and Nanodevices Simulation Laboratory, Department of Physics and Nanotechnology, Faculty of Engineering and Technology, SRM Institute of Science and Technology, Kattankulathur, Tamil Nadu, India

Muhammdh Sahaafi
Department of Chemistry, Periyar Maniammai Institute of Science & Technology, Thanjavur, Tamil Nadu, India

Khurshed Ahmad Shah
PG Department of Physics, Sri Pratap College, School of Sciences, Cluster University Srinagar, Srinagar, Jammu and Kashmir, India

S. Subash Chandra Bose
Centre for Functionalized Materials (CFMs), Department of Physics, PRIST Deemed University, Thanjavur, Tamil Nadu, India

D.M. Tejashwini
Department of Studies in Chemistry, Davangere University, Shivagangothri, Davangere, Karnataka, India

John David Thiruvadigal
Computational Materials Science and Nanodevices Simulation Laboratory, Department of Physics and Nanotechnology, Faculty of Engineering and

Technology, SRM Institute of Science and Technology, Kattankulathur, Tamil Nadu, India

Roshini K. Thumpakara
Carmel College (Autonomous), Mala Thrissur Kerala, India – 680732

A. Yousuf
PG Department of Physics, Sri Pratap College, School of Sciences, Cluster University Srinagar, Srinagar, Jammu and Kashmir, India

Chapter 1

Introductory aspects of nanoelectronics and applications

Khurshed Ahmad Shah, Ulfat T. Mir, and Brijesh Kumbhani

1.1 INTRODUCTORY ASPECTS

1.1.1 Free particle

The simplest system in quantum mechanics is one where the potential energy $V = 0$ everywhere. This is referred to as a free particle as it experiences no forces. In the one-dimensional case, where motion is restricted to the x-direction, the free particle is represented by a plane wave:

$$\psi = e^{ikx} \tag{1.1}$$

The time-independent Schrödinger equation for a free particle is given by:

$$-\frac{\hbar^2}{2m}\frac{d^2\psi}{dx^2} = E\psi$$

$$\frac{d^2\psi}{dx^2} = -\frac{2mE\psi}{\hbar^2} \tag{1.2}$$

Let $k^2 = \frac{2mE}{\hbar^2}$ where k is called wave vector, which is related to the particle momentum by $p = \hbar k$. Also, $\lambda = \frac{h}{p}$ can be written as $p = \frac{h}{\lambda} = \frac{h}{\lambda} \cdot \frac{2\pi}{2\pi} = \frac{\hbar}{\frac{\lambda}{2\pi}} = \hbar k$.

Substituting for k^2 in (1.2), we get:

$$\frac{d^2\psi}{dx^2} + \kappa^2\psi = 0 \tag{1.3}$$

DOI: 10.1201/9781003512899-1

Equation (1.3) has a general solution involving linear combination of plane waves:

$$\psi = Ae^{ikx} + Be^{-ikx} \tag{1.4}$$

where A and B are constant coefficients to be determined.

As a plane wave propagates, it repeats and, thus, the only constraints on it is periodicity. That is, it has periodic boundary conditions, also called Bon Van Boundary conditions:

$$\psi(x) = \psi(x + \lambda)$$

where λ is the wavelength or period of the wave

$$Ae^{ikx} = Ae^{ik(x+\lambda)} = e^{ikx}.e^{ik\lambda} \tag{1.5}$$

A similar equality applies to the second term. Thus, for either to be true $e^{ik\lambda} = 1$ as Euler relation states:

$$e^{ik\lambda} = \cos\ \kappa\lambda + i \sin \kappa\lambda$$

$$\kappa\lambda = 2\pi n$$

$$\kappa = \frac{2\pi n}{\lambda}$$

as constraints on κ that arises due to periodic boundary conditions of the system.

As there are no constraints on A or B, we are free to choose solutions. Let $B = 0$, the equation (1.4) becomes:

$$\psi = Ae^{ikx} \tag{1.6}$$

The accompanying energy is found by using the definition of wave vector, $k = \sqrt{\frac{2mE}{\hbar^2}}$.

This reveals a parabolic energy dependence on k.

1.1.2 Bohr radius

The Bohr radius is a physical constant that represents the approximate distance most likely to be found between the nucleus and the electron in the ground state of a hydrogen atom (stationary state of lowest energy). It is denoted by the symbol a_o and is named after the famous physicist Neils Bohr.

According to Bohr's theory, electrons orbit the nucleus in specific, fixed paths corresponding to discrete energy levels. Bohr suggested that electrons can only occupy these quantized energy states around the atom. Classical electromagnetic theory, however, predicts that an accelerating charged particle, such as an electron, would emit radiation in the form of electromagnetic waves. This energy loss should cause the electron to spiral inward toward the nucleus. As the electron moves inward, its angular velocity and frequency would vary, leading to the emission of light with continuously changing frequencies. This would result in a continuous spectrum, which contradicts the observed discrete line spectrum.

This led Bohr to conclude that the electron's angular momentum must be quantized, specifically an integer multiple of Planck's constant (6.626 × 10^{-34} m^2 kg) divided by 2π. This insight aligned with the observed discrete energy levels in emission spectra and allowed for calculating a fixed radius for each energy level. For the simplest atom, hydrogen, the single electron orbits the nucleus, with its lowest energy state corresponding to an orbital radius very close to the Bohr radius. However, due to the reduced mass effect, this radius is slightly smaller, differing by approximately 0.05%.

The Bohr model of the atom was eventually replaced by the concept of an electron probability cloud described by the Schrödinger equation, published in 1926. This framework, further refined by incorporating spin and quantum vacuum effects, accounts for fine structure and hyperfine structure. Despite this, the Bohr radius formula remains a cornerstone in atomic physics due to its straightforward relationship with fundamental constants. For this reason, it is defined using the true electron mass rather than the reduced mass, as previously noted. The Bohr radius has also become a standard unit of length in atomic units. In Schrödinger's quantum mechanical model of the hydrogen atom, the Bohr radius corresponds to the radial coordinate where the radial probability density of the electron's position is highest. However, the expected (average) radial distance of the electron is slightly different. The value of the Bohr radius is 5.2917721067 × 10^{-11} m, or 0.53 Å, which is also the radius of the first Bohr orbit. Bohr radius is given by:

$$a_o = \frac{4\pi\epsilon_o \hbar^2}{e^2 m_e} = \frac{\hbar}{m_e c\alpha} \tag{1.7}$$

where a_o is Bohr radius, m_e is the mass of electron at rest, ϵ_o is permittivity of free space, h is reduced Planck's constant, c is the velocity of light, α is the fine structure constant, and e is the elementary charge on the particle.

1.1.3 Excitons

An exciton is a bound state of an electron and a hole that are attracted to each other by the Coulomb force. Excitons are electrically neutral quasiparticles that can move in a coherent, particle-like manner in a material. They are found in many materials, including insulators, semiconductors, some metals, atoms, molecules, and liquids. As the electron and the positive hole carry equal but opposite electrical charges, the exciton as a whole is electrically neutral, though it carries energy. This neutrality makes excitons challenging to detect directly, but they can be observed through indirect methods.

When an electron within an exciton recombines with a positive hole, the atom is restored, and the exciton ceases to exist. During this process, the exciton's energy can be either emitted as light or transferred to a neighboring atom's electron within the solid. If the energy is transferred, the neighboring electron is displaced from its atom, forming a new exciton. Excitons are classified into two main types: Frenkel excitons and Wannier–Mott excitons.

1.1.3.1 Frenkel excitons

Yakov Frenkel first introduced the concept of excitons, describing the excitation of atoms within the lattice of a specific excitonic insulator. Frenkel excitons are characterized by a relatively low dielectric constant and a binding energy ranging from 0.1 to 1 eV. This occurs because the Coulomb interaction between the electron and hole is often strong and significant. Due to this strong interaction, Frenkel excitons are compact, resulting in their low dielectric constant. They are commonly found in crystals and organic molecular crystals such as anthracene and tetracene. These excitons have a tightly bound radius comparable to atomic radii or the size of a crystal's unit cell. Frequently observed in organic semiconductors or insulators, Frenkel excitons are sometimes referred to as bound electrons.

1.1.3.2 Wannier–mott excitons

In semiconductors, the dielectric constant is typically high, which reduces the Coulomb interaction between charged particles by screening the electric field. This process leads to the formation of Wannier–Mott excitons. These excitons have radii significantly larger than the lattice spacing. The large radii are primarily due to the small effective masses of electrons, a common characteristic of semiconductors. In this scenario, the lattice potential is incorporated into the electron and hole masses, resulting in the formation of an exciton polariton. Wannier–Mott excitons have low binding energy, generally around 0.01 eV. They are commonly found in semiconductor crystals and liquids, such as xenon. Their radii exceed the size of the unit cell, and they are typical in inorganic semiconductors. In addition, their excitonic bonds can be easily broken at room temperature.

They are distinguished by E_H binding energies. Mott–Wannier excitons have weak Coulomb interactions and remain relatively distant apart each other. Binding energy is ≈10 MeV (unstable at room temperature ~25 MeV), while Frenkel exciton is strongly bound ≈100 MeV. So, Mott–Wannier-type exciton is delocalized, while Frenkel exciton is localized. Semiconductors usually have Mott–Wannierexitons because of their large dielectric constants. In contrast, Frenkel excitons are typically observed in organic materials with relatively low dielectric constants. Excitons can occasionally be detected in the absorption spectrum of bulk semiconductors. They usually appear just below the band edge transition, as their energy is slightly less than the semiconductor's bandgap by an amount equivalent to their binding energy. In particular, the exciton formation energy is:

$$E_x = E_g - E_{bind} \tag{1.8}$$

where E_g is semiconductor bandgap

E_{bind} is exciton binding energy

If $E_{bind} = 0$

$E_x = E_g$

If $E_{bind} = E_g$

$E_x = 0$ and so on.

As a result, excitons can readily dissociate into their components, an electron and a hole. Excitons are most noticeable in the absorption spectra of bulk semiconductors at low temperatures, where (kT) is approximately 1 meV. However, the behavior changes in nanostructures, as the physical size constraints of the system force the electron and hole to stay in close proximity to one another.

1.1.3.3 Exciton bohr radius

The exciton Bohr radius is the distance between an electron and a hole that are paired together. The Bohr radius plays a role in this by being able to calculate it for a hydrogen atom and then using the effective mass of the electron and hole to translate it to an exciton. In a quantum dot, the nanoparticle can be smaller than the exciton Bohr radius because of confinement, which results in enhanced Coulomb interaction and discrete energy states. This is due to the absence of center-of-mass motion [1].

The exciton Bohr radius is an important quantity in the study of semiconductor materials because it determines the exciton's binding energy and the range of its interactions with other particles. It also provides information about the material's electronic and optical properties, such as its absorption and emission spectra. The exciton Bohr radius varies between different materials due to differences in their electron and hole masses, as well as their relative permittivity. Materials with heavier particles and higher permittivity values tend to have larger exciton Bohr radii, whereas materials with lighter particles and lower permittivity values have smaller exciton Bohr radii. The exciton Bohr radius is an important factor in the design and development of optoelectronic devices, such as solar cells and light-emitting diodes (LEDs). It affects the efficiency and performance of these devices by determining the maximum distance over which electrons and holes can be separated and recombined, as well as the wavelength of light that can be absorbed or emitted. Understanding and controlling the exciton Bohr radius are crucial in improving the functionality and efficiency of these devices.

Consider a two-particle system. We know, for electron:

$$a_e = \frac{4\pi\epsilon\epsilon_o\hbar^2}{e^2 m_e}$$

And for hole:

$$a_h = \frac{4\pi\epsilon\epsilon_o\hbar^2}{e^2 m_e}$$

An expression for electron–hole pair called as exciton is same, except replacing m_e or m_h by m_{exc}, which is $\frac{1}{\mu} = \frac{1}{m_e} + \frac{1}{m_h}$. Therefore, bulk exciton Bohr radius is:

$$a_b = \frac{4\pi\epsilon\epsilon_o\hbar^2}{e^2\mu}$$

$$a_b = \frac{4\pi\epsilon\epsilon_o\hbar^2}{e^2}\left[\frac{1}{m_e} + \frac{1}{m_h}\right]$$

$$a_b = a_e + a_h$$

In terms of Bohr radius:

$$a_b = \frac{\epsilon\, m_o}{\mu} a_o \tag{1.9}$$

where ε is the dielectric constant, m is the mass, μ is the reduced mass, and a_b represents the Bohr radius (~0.053 nm).

When particle size approaches exciton Bohr radius, the quantum confinement effect (QCE) causes increasing of the excitonic transition energy and blue shift absorption and luminescence bandgap energy. In addition, QCE leads discreteness to continuous bandgap of bulk material like atomic energy levels, which in turn leads discrete absorption. Based on confinement, confined structure is classified as quantum well, quantum wire, quantum dot or nanocrystal.

Structure	*Quantum confinement*	*Number of free dimensional*
Bulk	0	3
Quantum well/super lattice	1	2
Quantum wire	2	1
Quantum/nanocrystal	3	0

1.1.4 Relation between Bohr radius and de Broglie wavelength

Electron Bohr radius is:

$$a_e = \frac{4\pi\, \epsilon\epsilon_o \hbar^2}{e^2 m_e} \tag{1.10}$$

The velocity of electron for $n = 1$ is:

$$\begin{aligned} v_e &= \frac{n\hbar}{mr} = \frac{1.\hbar}{m_e r_{BR}} = \frac{\hbar}{m_e a_e} \\ v_e &= \frac{\hbar}{m_e .4\pi\epsilon\epsilon_o \hbar^2} m_e e^2 \\ v_e &= \frac{ne^2}{4\pi\, \epsilon\epsilon_o \hbar} \end{aligned} \tag{1.11}$$

Now, wavelength associated with such electron is:

$$\begin{aligned}
\lambda_e &= \frac{h}{p} = \frac{h}{m_e v_e} = \frac{\hbar}{n m_e e^2} 4\pi \epsilon\epsilon_o \hbar \\
\lambda_e &= \frac{2\epsilon\epsilon_o h^2}{m_e e^2} \\
n\lambda_e &= 2\pi a_e
\end{aligned} \tag{1.12}$$

As $2\pi r$ represents the circumference of the Bohr orbit (r), the equation above demonstrates that the circumference of the Bohr orbit in a hydrogen atom is an integer multiple of the de Broglie wavelength associated with the electron orbiting within it.

1.1.5 Atoms, clusters, and nanomaterials

Oswald was the first to realize that nanoscale particles exhibit novel and interesting properties, largely dependent on their size and shape.

Clusters are inorganic particles composed of a few dozen to a few hundred atoms. This interest has extended to a wide range of metals and semiconductors, driven by the spherical properties displayed by these nanometer-sized particles. Semiconducting clusters are held together with strong directional covalent bonds. Metallic clusters are fairly strongly held together by the localized, non-directional bonding [2]. Atoms are the foundation of matter, clusters bridge the gap between atomic and bulk scales, and nanomaterials exploit these nanoscale phenomena for innovative applications. Each plays a crucial role in advancing our understanding of material properties and enabling cutting-edge technologies.

Difference between atoms, molecules, clusters, and bulk material

Property	*Atom/Molecule*	*Cluster*	*Bulk material*
Size of constituting entities	Few angstrom	Angstrom to nanometer	Micron to higher
Number of constituents	One for atom, many for molecule	Two to several thousand	Infinite
Electron structure	Confined quantized	Confined quantized	Continuous
Geometrical structure	Well-defined and predictable	Well-defined and predictable	Crystal structure decides
Examples	Na, NaCl C_6H_6	C_{60}, NaCl	Bulk, silver, gold.

1.1.6 De Broglie wavelength of an electron

Nanomaterial is the material of a dimension at which quantum size effects are observed. At that occasion, material lies in the range of the De Broglie wavelength of electron in that material or Bohr excite on radius or de Broglie wavelength of exciton bond state in weak confinement or Bohr radius of exciton. We need to see all of similar types.

Case 1 free electron (approximately a metallic case)

For a free electron equation for wavelength is:

$$\lambda = \frac{h}{p} = \frac{h}{mv}$$

$$\lambda = \frac{6.64 \times 10^{-34}}{9.1 \times 10^{-31} \times \frac{1}{100} \times 3 \times 10^{8}} = 0.24\text{nm} \qquad (1.13)$$

1.1.7 De Broglie wavelength of C_{60}

Fullerene C_{60}, also known as buckyballs or buckminsterfullerene, is a nanostructured allotrope of carbon. This spherical allotrope consists of 60 carbon atoms arranged in a pattern of pentagons and hexagons, resembling the geometry of a soccer ball. The carbon atoms are bonded through alternating single and double bonds, giving rise to the term "buckyballs." Their cage-like structure and polyaromatic nature result in a displaced electron cloud, enabling these molecules to function as charge-transfer complexes. The diameter of a buckyball spans only a few atoms.

$$m_{60} = \frac{12.011}{N_a \times 1000} \times 60 = 1.2 \times 10^{-24}\,\text{kg} \qquad (1.14)$$

λ_{C60} = 2.5 pm, which is smaller than free electron.

For this reason, C60 is treated as a particle rather than a wave.

1.1.8 De Broglie wavelength of electron in semiconductor

The effective mass of electron in semiconductors is a fraction of m_o (free electron mass) and velocity 1×10^5 m/s (non-relative).

$$\lambda = \frac{h}{m_e ff^v} \qquad (1.15)$$

$\lambda_e = 7.3 \times 10^{-8}$ m = 73 nm

Comparing

$\lambda_{e\text{-}free} = 0.24$ nm

$\lambda_{e\text{-}semi}$= 73 nm suggests that a semiconductor can be nanomaterial at ~73 nm range, while as metal still tends to show bulk like signature down to 73 nm till sum 0.5 nm.

1.1.9 De Broglie wavelength of hole in a semiconductor

Hole is much heavier than electron in semiconductors, but both are lighter than free electron.

$$m_h = 0.4m_o \quad \text{(for non-relativistic hole)}$$

$$m_e = 0.1m_o$$

$$\lambda_h = 18.2 \text{ nm}$$

As per definition, a semiconductor material can behave as nanomaterial to electron species at large dimensions that of hole in the same.

1.2 INTERACTION OF LIGHT WITH MATTER (THE NANOPERSPECTIVE)

The interaction of light and matter at the nanoscale is swiftly gaining attention for its scientific and technological significance. Light–matter interactions at the nanoscale play a vital role for the efficient transformation of light into chemical energy in biological systems and into electrical current in artificial photovoltaic devices. These interactions contribute to the unique linear and nonlinear optical properties of metallic nanostructures, enabling the manipulation of nanoscale light localization through surface plasmon excitations [3]. This area, referred to as nanophotonics or nanooptics, centers on manipulating light–matter interactions at scales smaller than the diffraction limit. Photonics differs from traditional optics, which focuses on concepts like light rays and phenomena such as absorption, reflection, and transmission. Instead, photonics involves the light science study, including its emission, amplification, and detection.

In the past decade, with the rapid advancements in computational characterization techniques and advanced fabrication methods, researchers have conducted in-depth studies of light–matter interactions in the visible and near-infrared regions for nanostructures with dimensions on the order of, or smaller than, the wavelength.

Metal nanoparticles have been widely investigated for their ability to close the divide between traditional optics and advanced, highly integrated nanophotonic components by generating surface plasmons, which are oscillations of free electrons on their surface.

The negligible resistive losses in dielectric and semiconductor nanostructures enable the excitation of light at high intensities, allowing for significant light concentration at the nanoscale beyond just interfaces. These nanostructures exhibit multipolar electric and magnetic resonant modes, aiding in the engineering of light behavior at the nanoscale. In addition, the interaction of graphene with electromagnetic radiation is captivating due to its electron confinement and exceptional band structure. In summary, the field of photonics continues to expand, profoundly impacting various aspects of life through its advancements.

1.2.1 Optical properties of nanomaterials and size effect

The optical properties of nanomaterials, including reflection, transmission, absorption, and light emission, are intricately linked to their electronic structure, which is strongly influenced by their morphology. As the electronic structure of nanomaterials depends heavily on surface atoms, two main factors—quantum effects and surface plasmon resonance (SPR)—play a key role in determining the optical properties that vary with size. When the particle size is reduced to the nanoscale (around 100 nm), significant changes in optical properties are observed. Nanoparticles are so small that their electrons have limited freedom of movement compared to bulk materials. As a result, quantum confinement of electrons affects how nanoparticles interact with light, leading to distinct behaviors compared to bulk materials. The reduced dimensionality of nanomaterials greatly impacts the energies of the highest occupied molecular orbital (valence band or VB) and the lowest unoccupied molecular orbital (conduction band or CB). Optical phenomena like emission and absorption arise from electron transitions between different energy bands. The optical bandgap widens as the particle size decreases, particularly in semiconductor nanomaterials. By modifying the size and composition of nanomaterials, emission wavelengths can be adjusted across the UV, visible, and near-infrared spectra. For example, changing the size of CdS core–shell nanoparticles from 2 to 6 nm in diameter causes the emission wavelength to shift across the visible spectrum, with smaller particles emitting blue light and larger particles emitting red light. As the particle size decreases below the wavelength of the incoming radiation, SPR emerges as the dominant factor shaping the optical properties of nanomaterials. SPR occurs as electrons in the CB oscillate in phase with the applied light energy, producing distinct optical effects not seen in bulk materials. The intensity of SPR is directly linked to the quantity of excited electrons and the dielectric constant of the surrounding environment. Metals such as silver, aluminum, and gold are capable of supporting surface plasmon modes, where free electrons naturally oscillate at frequencies influenced by the particle's composition, size, and shape [4].

1.2.2 Surface plasmons

Surface plasmons are electromagnetic waves that propagate along the surface at the boundary between a metal and a dielectric (or metal/vacuum) medium. These excitations occur at the boundary where the electromagnetic waves propagate parallel to the interface between the metal and the dielectric material.

Surface plasmons are plasmons confined to surfaces that interact strongly with light, leading to the formation of polaritons. They occur at the interface between materials with positive relative permittivity, such as dielectrics (e.g., air, glass, vacuum), and materials with negative permittivity at a specific frequency of light, such as metals or heavily doped semiconductors. The magnitude of the negative permittivity region must be greater than the magnitude of the positive permittivity region for surface plasmons to exist; otherwise, light is not bound to the surface. At visible wavelengths, surface plasmon-supporting interfaces are commonly formed by metals such as silver or gold in contact with dielectrics, such as air or silicon dioxide.

Surface plasmons play a key role in surface-enhanced Raman spectroscopy, offering biochemists a tool to investigate the mechanisms and kinetics of ligand binding to receptors. They can also be observed in X-ray emission spectra of metals. Furthermore, surface plasmons are used to manipulate the color of materials by affecting how light interacts with the surface.

1.2.2.1 Surface plasmon resonance

SPR occurs when polarized light hits a thin metal film under conditions of total internal reflection. When a light beam hits a half-circular prism, it bends toward the interface plane as it moves from a denser medium to a less dense one. By adjusting the incident angle θ, the outgoing light changes until it reaches a critical angle, at which point all incoming light is reflected within the circular prism. This is referred to as total internal reflection.

In total internal reflection, although no light exits the prism, the electric field of photons extends approximately a quarter of a wavelength beyond the reflecting surface. When a thin film of noble metal, typically gold, is coated on the reflection side, it facilitates SPR. Gold is often used due to its convenient combination of reflectance angle and wavelength for SPR signals. When the energy of incident photons aligns correctly, they interact with the free electron clouds on the gold surface—specifically the outer shell and CB electrons. The energy of the absorbed photons is transferred to these electrons, generating surface plasmons according to quantum theory. A plasmon represents a particle-like electron density wave. During SPR, when the quantum energy of photons is appropriate, the photons are converted into plasmons, causing a decrease in reflected light intensity. SPR occurs when polarized light interacts with a metal film at the interface between media with different refractive indices. The SPR method induces collective oscillations of free electrons, called surface plasmons, using the Kretschmann configuration. In this arrangement, light is directed onto a metal film via a glass prism, and the resulting reflection is measured.

At a particular incident angle, called the resonance angle, plasmons resonate with light, causing the absorption of light at that angle, which produces a dark line in the reflected beam, and holds important information. The resonance angle can be determined by noting a dip in the intensity of SPR reflection. A shift in the reflectivity curve indicates a molecular binding event occurring on or near the metal film or bound to it. By tracking this shift over time, researchers can investigate binding kinetics and molecular binding events without the requirement for labels.

1.2.3 Nanophotonic and plasmonic applications

1.2.3.1 Nanolasers

As one of the most significant inventions of the 20th century, lasers have profoundly impacted human production and everyday life. Over the past 60 years, lasers have advanced rapidly in terms of smaller size, higher power, greater efficiency, and faster modulation speeds. The rapid development of nanomaterials and nanotechnology has significantly influenced the field of information and materials, providing the necessary conditions for the miniaturization of lasers. However, conventional lasers are limited by the diffraction limit, meaning their size cannot be smaller than half the wavelength, which poses challenges to further miniaturization. By the dawn of the new millennium, optically pumped photonic crystal nanolasers emerged as some of the smallest devices. Despite these advancements, issues such as reduced resonator size, low mechanical stability, ineffective heat conduction and scattering, and absorption losses have arisen, particularly due to the use of thin suspended membranes made of semiconductor material in their construction. The efficient miniaturization of lasers was significantly advanced with the development of metal-based cavities operating in dielectric or plasmonic modes. These devices use the distinct properties of metals to meet key laser requirements, such as thermal management, electrical contacts, optical confinement, and feedback. This led to the emergence of surface plasmonic nanolasers, or spasers (surface plasma amplification by stimulated emission of radiation). These devices use metals to support surface electromagnetic waves, which rely on electron oscillations to enable optical confinement. Stimulated emissions can amplify surface plasmons, with the gain medium usually consisting of dye molecules, semiconductor quantum dots, or bulk semiconductors. An external source injects holes into the VB of the gain medium and electrons into the CB, resulting in a population inversion. Given the high free carrier density required, the formation of excitons is unlikely. The transitions between the CB and VB in the gain medium are coupled to surface plasmon excitation/annihilation transitions via near-field radiation less interactions. Nanolasers exhibit unique and general properties due to their deeply sub-wavelength quasi-static nature, where traditional electromagnetic scales (wavelength, decay length, skin depth, etc.) become irrelevant. Their behavior depends primarily on the size of the metal nanostructure. Nanolasers have distinct properties, setting them apart from micro- or macroscopic lasers. The condition for nanolasing action does not rely on the size of the surface plasmon in the nanolaser and is determined by the following inequality:

$$\frac{4\pi}{3}\frac{|d|^2 n_e Q}{\hbar\Gamma_{12}\epsilon_d} \geq 1 \tag{1.16}$$

where n_e refers to the volume density of active electrons within the gain medium, Q is the metal quality factor, T_{12} is the spectral width of the nanolaser surface plasmon transition, ϵ_d represents the dielectric permittivity of the surrounding medium, and d is the electronic transition dipole in gain medium.

The nanolasing frequency w_s is a linear average of the gain transition frequency. w_{21} and surface plasmon frequency w_p:

$$w_s = \frac{\gamma_p w_{21} + T_{12}\, w_p}{\gamma_p + T_{12}} \tag{1.17}$$

where γ_p is the SPR width. This is similar to the frequency pulling effect in traditional lasers with detuning. Instead of the cavity frequency, the plasmon frequency is used.

For a surface plasmon nanolaser, the spontaneous decay rate γ_2 of the gain medium excitation into the surface plasmon lasing mode is plasmonically enhanced and can be expressed for a nano shell laser as follows:

$$\gamma_2 = \frac{2|d|^2}{\hbar\varepsilon_d \varepsilon'_m \varepsilon'_{m'} R^3} \tag{1.18}$$

where R is the nano shell radius.

Above the surface plasmon amplification(spacing), threshold, surface plasmon is stimulated and its emission rate can be expected to rise with the increase in pumping intensity.

Nanolasers have the potential for a wide range of applications across various fields. They can be used in biosensors, microscopy, laser surgery, chemical and biomedical engineering, material processing, visualizing single virus or DNA molecules, nano-switching, information storage, and nanoanalysis [5].

1.2.3.2 Nanoantenna

A nano antenna is an antenna, which is extremely small and is not used to carry sound waves. It is used to gain understanding about the interactions happening at the nanoscale. The theory behind nano antennas is similar to that of rectifying antennas. When light hits the antenna, the electrons within the antenna oscillate back and forth at the same frequency as the incoming light, driven by the oscillating electric field of the electromagnetic wave. Nano antennas are crucial in various fields, including space communication, optical wireless communication, mobile communication, radar object detection, wireless data transfer, and nanophotonic applications.

The fabrication of nano antennas can be done using electron beam lithography, focused iron beam lithography, and nano imprinting lithography. Nano imprinting lithography is the best method of fabrication as it is low cost, less time-consuming, and high-throughput technique. The material selection in nano antenna fabrication is a difficult problem. Currently, gold and silver are being used as the material for nano antenna fabrication; however, they exhibit skin effect at higher frequencies, which affects the efficiency of the nano antennas. To eliminate this problem, nano antennas based on materials like carbon nanotubes and graphene have been realized.

1.2.3.3 Photonic crystals

Photonic crystals are macroporous materials with unique and interesting optical properties. They consist of a periodic arrangement of matter on a scale roughly corresponding to the wavelength of visible light, approximately several hundred nanometers. As photonic crystals are structured on a length scale similar to the influence of light, they affect light propagation. Although X-ray diffraction does not provide sufficient information about the structure of photonic crystals, neutron diffraction is a more suitable technique for analysis. In addition, light diffraction can be used for studying photonic crystals. However, due to the strong interaction between visible light and photonic crystals, multiple scattering occurs, making the elucidation of their structure quite challenging [6].

1.2.4 Electroluminescence

Electroluminescence is an optical phenomenon in which photons are emitted due to the radiative recombination of electrons and holes as an electric current passes through a material. Electroluminescence is broadly classified into two types: low-field electroluminescence and high-field electroluminescence. In low-field electroluminescence, minority charge carriers are injected, whereas in high-field electroluminescence, majority charge carriers are accelerated to high optical energy levels. The low-field type is commonly seen in PN junction semiconductors like LEDs, whereas the high-field type is prevalent in electroluminescent displays composed of organic luminescent materials or phosphors.

1.2.5 Photochromic nanomaterials

Photochromism refers to the reversible changes in a material's absorption spectrum when exposed to ultraviolet (UV) light. These materials undergo a color change that is reversible with variations in UV light intensity. In the absence of UV light, photochromic materials are typically colorless, but as UV intensity increases, they become colored. The materials revert to their colorless state as UV light intensity decreases. Incorporating photoresponsive units into nanomaterials enhances their ability to activate specific functions with higher precision and selectivity. Rotaxane, for example, is a photochromic nanomaterial used in molecular machines. Various functional nanoparticles, such as noble metal nanoparticles, mesoporous silica nanoparticles, fluorescent nanoparticles, and magnetic nanoparticles, exhibit photochromism, offering numerous applications. Photochromism is also observed in molecular junctions, organic field-effect transistors, and other areas, enabling the direct observation of novel effects and the discovery of fundamental physical phenomena beyond traditional materials [7].

1.2.6 Electrochromic nanomaterials

Electrochromism is described as the change in materials optical properties under an applied voltage. It involves the ions, insertion, and extraction in to/out of electrochromic material. Nanostructures, characterized by their small dimensions and large specific surface areas, are engineered to enhance the iron insertion and extraction process, thus improving the performance of electrochromic materials and devices. Some of the examples of electrochromic nanomaterials are ZnO nanorods, crystalline WO_3 nanoparticles and nanorods, mesoporous TiO_2, mesoporous WO_3, V_2O_5 nanowires, Prussian Blue Pb, nano inks, and nanostructures.

Nanostructures with extensive specific surface areas enable rapid and stable electrochromic switching. These nanostructures may include ultrathin crystalline nanorods, nanotubes, nanowires, quantum dots, and mesoporous crystalline structures.

1.3 CONCLUSION

Nanophysics serves as a vital link between the quantum realm and observable phenomena, revealing the distinct properties of matter at the nanoscale. This chapter highlighted key concepts such as the quantum behavior of particles, the significance of the Bohr radius, and the interplay between wave–particle duality and nanoscale systems. These ideas underscore the importance of quantum effects in understanding and designing nanosystems, from individual atoms to complex nanomaterials. The interaction of light with nanomaterials was also explored, emphasizing how size and structure influence optical properties. Phenomena like SPR and their applications in nanotechnology, including nanolasers, photonic crystals, and chromic materials, demonstrate the transformative potential of nanophysics in advancing technology.

Overall, the study of nanophysics provides a deeper understanding of the unique behaviors and interactions at the nanoscale, offering a foundation for scientific innovation and the development of groundbreaking technologies. This exploration opens the door to limitless opportunities in fields ranging from material science to optoelectronics.

1.4 FUTURE PROSPECTIVES

The concepts in this chapter lay the groundwork for advancements in nanotechnology. Innovations like nanophotonic devices, efficient plasmonic systems, and chromic materials hold potential for applications in telecommunications, smart devices, and sustainable technologies. Advances in SPR and related technologies could revolutionize fields like medical diagnostics, renewable energy, and environmental monitoring. In addition, the development of photochromic and electrochromic nanomaterials holds promise for smart windows, energy-efficient displays, and adaptive optical devices. As research continues to uncover the unique properties of nanomaterials, these insights will drive innovation in sustainability, quantum computing, and advanced manufacturing, shaping the future of science and technology.

REFERENCES

1. Khanna, V. K. (2020). *Introductory nanoelectronics: physical theory and device analysis*. CRC Press.
2. Madkour, L. H. (2019). *Nanoelectronic materials: fundamentals and applications* (Vol. 116). Springer.
3. Lienau, C., Noginov, M. A., Lončar, M., Bauer, C., & Giessen, H. (2014). Light–matter interactions at the nanoscale. *Journal of Optics*, *16*(110201), 110201.

4. Asha, A. B., & Narain, R. (2020). Nanomaterials properties. In *Polymer science and nanotechnology* (pp. 343–359). Elsevier.
5. Azzam, S. I., Kildishev, A. V., Ma, R. M., Ning, C. Z., Oulton, R., Shalaev, V. M., ... & Zhang, X. (2020). Ten years of spasers and plasmonicnanolasers. *Light: Science & Applications*, 9(1), 90.
6. Schmidt, W., Schüth, F., & Weidenthaler, C. (2013). Diffraction and spectroscopy of porous solids. In *Comprehensive inorganic chemistry II (second edition). From elements to applications; Volume 5: Porous materials and nanomaterials* (pp. 1–24). Elsevier.
7. Wang, L., & Li, Q. (2018). Photochromism into nanosystems: towards lighting up the future nanoworld. *Chemical Society Reviews*, 47(3), 1044–1097.

Chapter 2

Nanoelectronics

Exploring the building blocks of future technologies

Sheenu Bhadauria and Lal Kumar Chandel

2.1 INTRODUCTION

Modern technology places a strong emphasis on miniaturization, particularly evident in the field of electronics [13]. This has dramatically increased the number that can fit onto a single chip. Modern microprocessors now contain billions of transistors [4], enabling advanced computational capabilities. Memory chips, such as Dynamic Random Access Memory (DRAM) and NOT AND (NAND) flash, have increased storage density, allowing for multiple gigabits of data storage [5]. Application-specific integrated circuits (ASICs) and field-programmable gate arrays (FPGAs) serve as examples of the trend toward customization and the integration of multiple functions in small designs [6]. System on Chips (SoCs) combines central processing unit (CPUs), graphics processing unit (GPUs), and memory, optimizing space in mobile devices [7]. Mixed-signal and analog integrated circuits (ICs) integrate both analog and digital functions, which are crucial for communications and signal processing [8, 9]. As nanoelectronics progresses, it further increases transistor density and these innovations highlight the critical role of microelectronics in fostering advancements across diverse sectors, such as consumer electronics, telecommunications, automotive, and healthcare, paving the way for more efficient and powerful electronic devices, including consumer electronics, telecommunications, automotive, and healthcare [10–13].

This progression has empowered designers to craft more intricate ICs. Over the past three and a half decades, engineers have escalated the complexity of these circuits by over five orders of magnitude, fundamentally reshaping society. Even traditionally mechanical devices have increasingly integrated electronic components to enhance functionality, efficiency, and user experience. For instance, modern automobiles utilize engine control units (ECUs) for precise engine management [14], while washing machines incorporate microcontrollers for automated cycle control [15]. Refrigerators and air conditioning systems use thermostats and inverter circuits to optimize energy usage and maintain desired temperatures [16]. Digital cameras

DOI: 10.1201/9781003512899-2

rely on electronic sensors like charge-coupled device (CCDs) for image capture [17], and elevators utilize control panels for safety and automation [18]. Electric bicycles use battery management systems (BMS) to regulate speed [19], while robotic arms use servo motors for precision control [20]. Electric musical instruments like guitars and keyboards integrate audio interfaces, and heating systems benefit from programmable thermostats for efficient temperature regulation [21, 22]. These advancements highlight the synergy between mechanical and electronic components, leading to smarter and more efficient devices across various applications [23–29].

In parallel, nanoelectronics, a subfield within nanotechnology, focuses on harnessing nanoscale materials and techniques to forge electronic components. Operating at dimensions typically below 100 nanometers, these devices are pivotal to advancing fields such as computing. Nanoelectronics capitalize on innovative materials such as carbon nanotubes (CNTs), graphene [30], and semiconductor nanowires [31], each offering unique properties that enhance performance across various applications. These materials not only enable higher speeds and lower power consumption but also open new avenues in sectors such as telecommunications, computing, healthcare, and renewable energy [20–23]. The convergence of nanoelectronics with fields like quantum computing and nanophotonics further propels technological advancement, fostering interdisciplinary research and economic growth. Although challenges such as scaling limitations and manufacturing complexities persist, the boom in nanoelectronics continues to drive innovation, promising solutions to global challenges and reshaping the landscape of modern technology [32–35].

Over the years, reflecting advances in materials science, device physics, and computing paradigms nanoelectronics, marked by pivotal milestones that have shaped its historical context and trajectory. In 1958, Jack Kilby pioneered the IC at Texas Instruments, marking a transformative milestone in electronics. This innovation consolidated numerous components onto a single chip, catalyzing the shift toward miniaturized technology and paving the way for the rise of nanoscale electronics. It revolutionized the industry, enabling exponential advancements in computational power and device sophistication over the following decades

In the 1970s, Gordon Moore's observation, widely recognized as Moore's Law, outlined the doubling of transistors on ICs approximately every two years [36]. This insight became a cornerstone for the semiconductor industry, driving continuous innovation in nanoelectronics and fostering breakthroughs in materials science and fabrication techniques to develop smaller, more efficient devices.

Moore's Law has driven rapid technological progress, fundamentally transforming the modern world through its influence on computing power and electronic devices. Despite facing formidable challenges such as quantum mechanical effects and heat dissipation at nanoscale dimensions, the essence of innovation and progress embedded in Moore's Law continues to inspire new methodologies and technologies designed to sustain the momentum of technological advancement [37]. Sustaining this progression demands a comprehensive exploration of innovative materials such as graphene and transition metal dichalcogenides (TMDs), alongside cutting-edge device designs including spintronics, quantum dots (QDs), and single-electron transistors.

These innovations are pivotal as traditional scaling approaches encounter physical limitations. Nanoelectronics aim to extend the leveraging of quantum confinement, tunneling, and superposition phenomena to develop devices that transcend conventional silicon-based technologies. As the industry approaches the end of classical scaling, these novel approaches promise to uphold the trajectory of exponential growth in computational capabilities. Addressing these challenges and harnessing the potential of next-generation nanoelectronic devices will be crucial for navigating the future of computing and its broader societal impacts, ensuring the continued evolution of technology in the face of physical and theoretical constraints.

The 1980s marked the emergence of nanoelectromechanical systems (NEMS), integrating electrical and mechanical functionalities at the nanoscale. This interdisciplinary approach broadened the scope of nanoelectronics beyond traditional semiconductor devices, creating new paradigms for sensing, actuation, and communication [38]. The 1990s marked a pivotal era in nanomaterial research, highlighted by breakthroughs in the synthesis and analysis of semiconductor nanocrystals and QDs. These nanoscale entities showcased distinctive electronic and optical behaviors, sparking considerable interest for their application in highly compact electronic and optoelectronic technologies [39].

The nanoelectronics had become a cornerstone of nanotechnology. Poole Jr. and Owens provided a comprehensive introduction to nanotechnology, focusing on manipulating and controlling electronic properties at the nanoscale to achieve novel device functionalities and applications. Concurrently, there was a growing understanding of quantum mechanical effects in nanoscale devices, highlighted in the research works of semiconductor devices [40, 41].

In the 21st century, the exploration of quantum transport phenomena and quantum computing significantly expanded the horizons of nanoelectronics. Insights from Datta on quantum transport and Blick's exploration of quantum nanoelectronics exemplify the field's progression toward harnessing quantum effects for next-generation computing and communication technologies [42, 43].

To inspire innovation in nanoelectronics as we transition into an era where classical scaling reaches its physical limits, the field's ability to adapt and innovate with new materials and quantum principles will be paramount. This will not only sustain technological growth but also revolutionize the landscape of computing and its impact on society. Addressing the challenges and harnessing the potential of next-generation nanoelectronic devices will be crucial for navigating the future of computing and its broader societal impacts. The integration of interdisciplinary approaches and the exploration of novel materials and device architectures will ensure the continued evolution of technology in the face of physical and theoretical constraints, paving the way for a new era of technological advancement [44–56].

2.2 QUANTUM PHENOMENA IN NANOELECTRONICS

At the nanoscale, the behavior of electronic components is governed by quantum effects, giving rise to phenomena like tunneling and quantized conductance [57]. This section delves into the core principles of quantum mechanics and examines their impact on the field of nanoelectronics.

2.2.1 Quantum mechanics: The foundation of nanoelectronics

2.2.1.1 Wave–particle duality

Central to quantum mechanics is the principle of wave–particle duality, which describes how particles like electrons can display characteristics of both waves and particles. This phenomenon becomes especially important in nanoelectronics, where the dimensions of components approach the de Broglie wavelength of electrons.

This leads to observable quantum effects that are absent in larger, classical systems. Like, in a QD, an electron behaves as a standing wave confined within the boundaries of the dot, leading to discrete energy levels, much like the quantized energy levels in an atom. These energy levels are exploited in applications like QD lasers and single-electron transistors [58–61].

2.2.1.2 Heisenberg's uncertainty principle

The uncertainty principle asserts that the exact position and momentum of a particle cannot be determined simultaneously. This concept plays a critical role in nanoelectronics, where maintaining precise control over electron behavior is essential for optimal device functionality. This principle has profound implications in nanoelectronics, where the precise control of electron position and momentum is crucial for device operation. In tunneling phenomena, the uncertainty in an electron's position allows it to "tunnel" through potential barriers that would be insurmountable in classical physics [62–64].

2.2.2 Quantum tunneling: The path through barriers

Quantum tunneling is a remarkable phenomenon in quantum mechanics, where particles traverse potential energy barriers that would be insurmountable according to classical physics.

2.2.2.1 Tunneling in nanoelectronicdevices

In nanoelectronics, tunneling is not just a theoretical curiosity; it is a fundamental operation mechanism for several devices. The small dimensions of nanodevices often lead to situations where electrons can tunnel through insulating barriers, enabling or enhancing device functionality. For example, the scanning tunneling microscope (STM) utilizes the principle of quantum tunneling to achieve atomic-level imaging of surfaces. By detecting the tunneling current that flows when a sharp metallic tip is positioned extremely close to a sample, separated by a narrow insulating gap, the STM enables precise surface characterization [65–67].

2.2.2.2 Tunnel diodes

Tunnel diodes are a practical application of quantum tunneling in electronics. These diodes exhibit a region of negative differential resistance due to the quantum tunneling of charge carriers across a very narrow p–n junction. Because tunnel diodes may function at very high frequencies due to their quick tunneling process, they are used in high-speed switching circuits and oscillators [68].

2.2.3 Quantized conductance: Discreteness in conduction

In nanoelectronics, when the dimensions of the conducting channel approach the electron's wavelength, conductance becomes quantified, meaning the flow of electrons is restricted to discrete values rather than continuous.

2.2.3.1 Conductance quantization in nanowires

In a nanowire or a quantum point contact, the conductance occurs in discrete steps, with each step corresponding to a value of $2e^2/h$, where e is the electron charge and h is Planck's constant. This quantization occurs because the electron's wave function can only occupy discrete modes within the confined geometry of the nanowire. Likewise, when a nanowire is used to connect two large conductors, the conductance increases in discrete steps as the width of the wire is increased, corresponding to the addition of one more quantized conductance channel. This phenomenon is crucial in the design of quantum point contacts and has implications for the miniaturization of electronic components [69, 70].

A fundamental idea in the creation of quantum computing devices is quantized conductance, which states that qubits require quantum-level control over electron flow to function. For instance, the concepts of quantized conductance and tunneling are necessary to precisely regulate the motion of electrons in QDs, which are utilized as qubits in some kinds of quantum computers. This makes it possible to manipulate the quantum states that are required for quantum computing [71, 72].

2.2.4 Superposition and entanglement: Quantum states in nanoelectronics

Beyond tunneling and quantized conductance, two other quantum phenomena, superposition and entanglement, play crucial roles in the future of nanoelectronics, particularly in the development of quantum computing.

2.2.4.1 *Superposition in qubits*

In classical computing, a bit can only be in one of two states, 0 or 1, but in quantum computing, a qubit can simultaneously exist in a superposition of both states due to the principles of quantum mechanics. As an illustration, QDs can be designed to behave as qubits, using superposition to carry out intricate calculations more quickly than with traditional bits [73].

2.2.4.2 *Quantum entanglement in nanoelectronicdevices*

The phenomenon known as entanglement occurs when two or more particles' quantum states entangle, meaning that regardless of their distance from one another, in quantum mechanics, the state of one particle can instantly influence the state of another, a phenomenon known as entanglement. This property is leveraged in secure communication methods such as quantum key distribution (QKD), where entangled photons, generated in QD systems, ensure communication security [74,75].

2.3 NANOMATERIALS ENGINEERING IN NANOELECTRONICS

Nanomaterial engineering plays a pivotal role in nanoelectronics, enabling the design of electronic devices with customized properties by manipulating materials at the nanoscale. The distinctive features of nanomaterials, including their high surface area, quantum confinement effects, and superior electrical, optical, and mechanical attributes, make them highly suitable for a range of nanoelectronic applications. The key nanomaterials are used in nanoelectronics, their synthesis methods, and their application in real-world devices. Table 2.1 provides a comprehensive overview of various

Table 2.1 A comprehensive overview of various nanomaterials and their applications

Material category	*Preparation method*	*Particle size*	*Properties affected (increase/decrease)*	*Applications*	*Importance*	*Ref.*
Graphene	Mechanical exfoliation, CVD, liquid-phase exfoliation	1–10 nm (thickness) ~1 μm lateral size	*Increase:* Electrical conductivity, mechanical strength *Decrease:* Bandgap for semiconducting applications	Transparent conductors, sensors,energy storage devices	Excellent electrical and thermal conductivity, mechanical strength, used in flexible electronics and high-strength composites	[76]
CNTs	Arc discharge, laser ablation, CVD, floating catalyst, flame synthesis	1–100 nm (diameter) ~100 μm (length)	*Increase:* Electrical conductivity, tensile strength *Decrease:* Ductility at higher aspect ratios	Composite materials, sensors, drug delivery, nanoelectronics	High aspect ratio offers unique mechanical, thermal, and electrical properties, crucial for various applications	[77]
Nanowires (semiconductor nanomaterials)	Chemical vapor deposition, template synthesis	10–100 nm (diameter)	*Increase:* Electron mobility, quantum confinement effects *Decrease:* Mechanical strength at lower diameters	Nanoelectronics, optoelectronics, solar cells	High aspect ratio offers unique electrical and optical properties, crucial for next-gen transistors and sensors	[78]
QDs (semiconductor nanomaterials)	Colloidal synthesis, thermal decomposition	2–10 nm	*Increase:* Optical tunability, quantum confinement effects *Decrease:* Stability in bulk applications	Displays, solar cells, bioimaging, drug delivery	Size-dependent optical and electronic properties enable them for quantum computing, photovoltaics, and bioimaging	[79]

(*continued*)

Table 2.1 (Cont.)

Material category	*Preparation method*	*Particle size*	*Properties affected (increase/decrease)*	*Applications*	*Importance*	*Ref.*
2D materials: Transition metal dichalcogenides (TMDs)	Chemical vapor deposition, mechanical exfoliation	1–10 nm (thickness) ~1 μm lateral size	*Increase:* Direct bandgap in monolayers *Decrease:* Conductivity and mobility compared to graphene	Field-effect transistors, optoelectronics, catalysis	TMDs like MoS_2 exhibit direct bandgap in monolayer form, crucial for electronics and catalysis applications	[80]
2D materials: hexagonal boron nitride (hBN)	Chemical vapor deposition, mechanical exfoliation	1–10 nm (thickness) ~1 μm lateral size	*Increase:* Thermal stability, electrical insulation *Decrease:* Conductivity	Substrates for 2D materials, thermal interface materials, UV emitters	Excellent electrical insulation properties combined with high thermal stability and chemical resistance	[81]
Nanocomposites (e.g., CNT-polymer composites)	Various methods (mixing, in-situ polymerization)	*Variable:* Nanoparticles ~1–50 nm	*Increase:* Mechanical strength, conductivity *Decrease:* Flexibility in certain polymer matrices	Structural reinforcement, electronic packaging, energy storage	CNT and graphene-based composites enhance mechanical, thermal, and electrical properties of base materials	[82]
MOFs	Solvothermal synthesis, electrochemical methods	1–100 nm (crystal size)	*Increase:* Surface area, porosity *Decrease:* Mechanical stability for bulk applications	Gas storage, separation, catalysis, sensors	MOFs provide high surface area and tunable porosity, ideal for gas adsorption and catalysis	[83]
Nanocomposites (e.g., metal-polymer hybrids)	Various methods (mixing, in-situ polymerization)	*Variable:* Nanoparticles ~1–50 nm	*Increase:* Electrical conductivity, strength, wear resistance *Decrease:* Flexibility, ductility	Nanoparticle-based hybrid materials enhance multifunctionality, improving both mechanical and electrical properties	Electronics, sensors, catalysis, coatings.	[84]

nanomaterials, emphasizing their synthesis, size, properties, applications, and significance in this field.

2.3.1 Carbon-based nanomaterials

2.3.1.1 Carbon nanotubes

CNTs are cylindrical nanostructures formed by rolling up sheets of single-layer carbon atoms (graphene). They can be classified as single-walled carbon nanotubes (SWCNTs) or multi-walled carbon nanotubes (MWCNTs) (Figure 2.1), and are known for their exceptional electrical conductivity, mechanical strength, and thermal stability. CNTs play a key role in the development of field-effect transistors (CNTFETs), which have the potential to surpass traditional silicon-based transistors due to their higher electron mobility and ability to operate at the nanoscale. CNTFETs are being investigated for use in high-speed computing and flexible electronic devices [85]. CNTs can be prepared using several methods, each offering unique advantages represented in Table 2.2, which simplifies the details about each method, focusing on how each method influences particle size, its applications, and why each method is important.

2.3.1.2 Graphene

Graphene, as shown in Figure 2.1a and 2.1b, is a single layer of carbon atoms arranged in a hexagonal pattern, renowned for its outstanding electrical conductivity, mechanical strength, and flexibility. Its two-dimensional (2D) structure enables exceptional charge carrier mobility, making it a promising material for nanoelectronics. Graphene is utilized in high-frequency transistors, transparent conductive electrodes, and sensors. Notably, graphene-based transistors are being designed for flexible and wearable electronics, providing both superior performance and mechanical resilience [77].

2.3.2 Semiconductor nanomaterials

2.3.2.1 Semiconductor nanowires

Nanowires are 1D structures with diameters in the nanometer range and lengths extending up to several micrometers. Semiconductor nanowires, such as those composed of silicon, gallium arsenide (GaAs), or indium phosphide (InP), exhibit quantum confinement effects that significantly influence their electronic properties. Silicon nanowires are particularly useful in developing nanoscale transistors and sensors. These nanowires can be integrated into logic circuits, serving as a foundation for next-generation nanoelectronic devices that offer enhanced miniaturization, speed, and energy efficiency compared to traditional bulk materials [95–97].

Table 2.2 A summary of CNTs and their applications

Method name	*Particle size*	*Importance*	*Applications*	*References*
Arc discharge	1–10 nm diameter	Produces high-quality, multi-walled CNTs (MWCNTs) and single-walled CNTs (SWCNTs). Important for high-purity CNT production but limited by scalability.	High-strength composite materials, field emission devices, energy storage, and sensors	[77]
Laser ablation	1–2 nm diameter	High-quality SWCNTs production with low structural defects. Suitable for laboratory use due to high cost and low production rate.	Aerospace materials, conductive films, transistors, and high-frequency electronics	[86]
CVD	10–100 nm diameter	Versatile, scalable, and cost-effective. Suitable for large-scale production with control over the tube structure and alignment. Crucial for industrial-scale applications.	Flexible electronics, gas sensors, lithium-ion batteries, supercapacitors, and hydrogen storage	[87]
High-pressure carbon monoxide (HiPCO)	0.8–1.2 nm diameter	Produces high-purity SWCNTs, valuable for applications requiring precise control over diameter and chirality. Expensive due to the high-pressure requirement.	Field emission displays, biomedical applications (drug delivery), and nanoelectronics	[88]
Plasma-enhanced CVD (PECVD)	5–50 nm diameter	Enhanced vertical alignment of CNTs due to the plasma field. Important for device integration requiring specific alignment and structure.	Nanoelectronics, optoelectronic devices, and energy conversion systems	[89]
Floating catalyst method	20–100 nm diameter	Scalable method, particularly useful for producing continuous CNT fibers and sheets. A key method for industrial-scale production.	High-strength fibers, catalysts, and fuel cell components	[90]

Table 2.2 (Cont.)

Method name	*Particle size*	*Importance*	*Applications*	*References*
Solution-based methods (electrolysis)	30–50 nm diameter	Produces CNTs in liquid phases, valuable for dispersing CNTs in various matrices. However, quality control and scalability are challenging.	Coatings, conductive composites, and sensors	[91]
Ball milling	10–50 nm diameter	Economical and scalable method for producing CNTs from graphite. However, this method often leads to impurities and structural defects.	Structural reinforcement, conductive inks, and electronic materials	[92]
Flame synthesis	2–5 nm diameter	Simple and cost-effective method with relatively high production rates. However, the control over CNT structure and quality is limited.	Gas sensors, catalysis, and coatings	[93]
Template-based methods	10–100 nm diameter	Allows for the synthesis of CNTs with precise size and structure. The method is useful for specific applications requiring uniformity but is less scalable.	Nanostructured materials for electronics, drug delivery systems, and advanced filtration materials	[94]

2.3.2.2 *Quantum dots*

QDs are nanoscale semiconductor particles that display size-dependent optical and electronic properties resulting from quantum confinement. These characteristics make them perfect for applications requiring precise control of electron energy levels. QDs are commonly used in displays, such as QLED TVs, to produce vibrant and tunable colors. Furthermore, they are being investigated for use in quantum computing, where their discrete energy levels can function as qubits for complex computational tasks [98].

2.3.3 Two-dimensional materials

2.3.3.1 *Transition metal dichalcogenides*

TMDs, such as molybdenum disulfide (MoS_2) and tungsten diselenide (WSe_2), are layered materials known for their distinctive electronic and

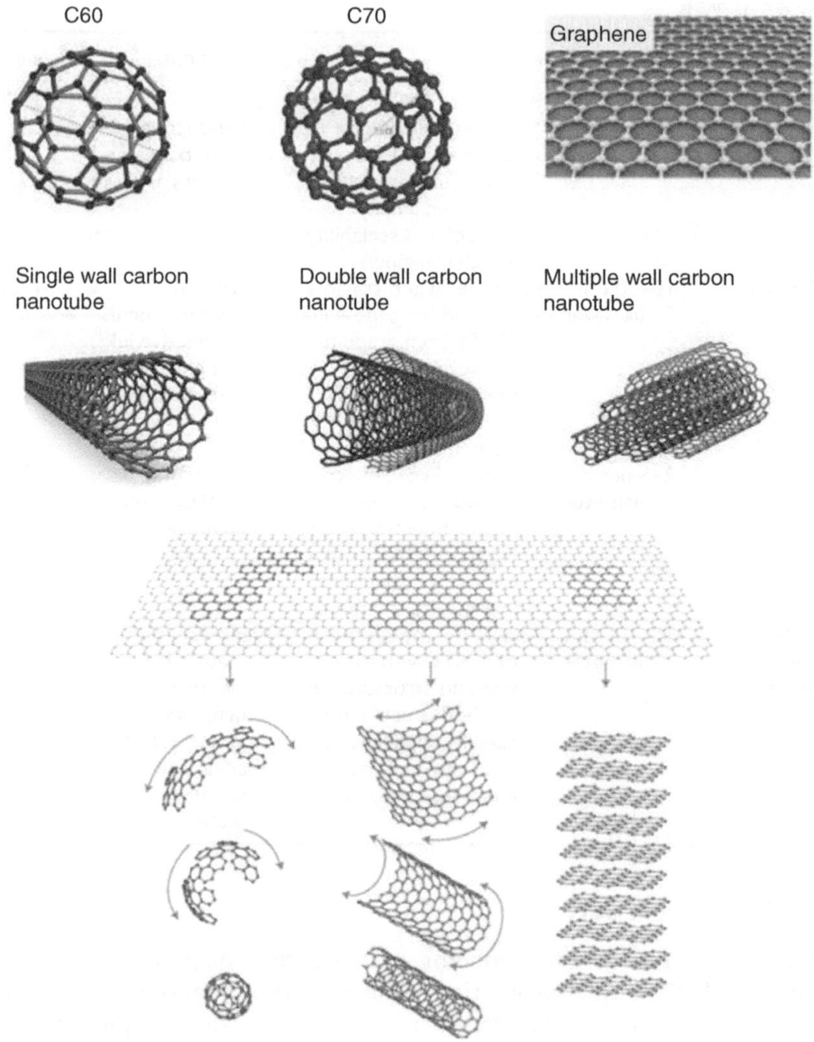

Figure 2.1 (a)The different CNTs and representation [77]. (b) The different CNTs and representation [45].

optical properties. These materials can be exfoliated into monolayers, similar to graphene, but exhibit semiconducting properties, making them ideal for various nanoelectronic applications. MoS_2 is particularly useful in the development of 2D transistors, which offer high on/off ratios and the ability to operate at low voltages. These transistors are well-suited for flexible electronics and low-power devices, helping to advance the miniaturization and energy efficiency of electronic systems [99].

2.3.3.2 *Hexagonal boron nitride*

Hexagonal boron nitride (h-BN) is a 2D material widely utilized as an insulating layer in nanoelectronic devices. With its large bandgap and exceptional thermal and chemical stability, it serves as an ideal substrate or dielectric material when combined with other 2D materials. h-BN is frequently used in graphene-based devices to create a smooth, insulating surface that improves the performance of graphene transistors. In addition, its role as a dielectric layer in 2D heterostructures aids in advancing nanoscale electronic and optoelectronic devices [100].

2.3.4 Nanocomposites and hybrid materials

2.3.4.1 *Metal-organic frameworks*

Metal-organic frameworks (MOFs) are porous materials made up of metal ions coordinated to organic ligands. Their adjustable porosity and capacity to incorporate diverse functional groups make them valuable in nanoelectronics, especially for sensors and storage devices. MOFs are being investigated for use in gas sensors, where their high surface area and customizable chemistry enable the selective detection of gases at extremely low concentrations. These sensors can be integrated into nanoelectronic systems for applications in environmental monitoring and industrial safety [101].

2.3.4.2 *Nanocomposites*

Nanocomposites are materials that consist of a polymeric matrix embedded with nanoparticles or nanofillers. By integrating the matrix's properties with the unique characteristics of the nanofillers, these materials offer enhanced mechanical, electrical, and thermal performance. Conductive nanocomposites, created by incorporating CNTs or graphene into a polymer matrix, are used in flexible electronics, electromagnetic interference (EMI) shielding, and wearable devices. These materials combine flexibility, conductivity, and durability, making them ideal for the development of advanced electronic devices [102,103].

2.4 FABRICATION TECHNIQUES IN NANOELECTRONICS

In nanoelectronics, particle size and method selection are crucial for device performance and functionality. The nanometer range, allow for enhanced electrical, thermal, and mechanical properties, essential for miniaturized. The method chosen to create these particles affects the material's quality, uniformity, and scalability. For example, chemical vapor deposition (CVD) ensures precise control over thin film formation, while mechanical methods may provide higher purity but limit large-scale production. Selecting the right method and particle size directly influences conductivity, efficiency, and integration into advanced nanoelectronic systems.

The fabrication of nanoelectronic devices requires advanced techniques capable of creating structures at the nanometer scale with high precision and efficiency. The development of innovative fabrication methods becomes crucial. An overview of traditional and emerging fabrication techniques, and highlights their role in the realization of cutting-edge nanoelectronic devices.

The creation of nanoelectronic devices necessitates advanced fabrication techniques that can construct structures at the nanometer scale with exceptional precision and efficiency. As the need for smaller, faster, and more energy-efficient devices grows, the evolution of novel fabrication methods becomes essential. This includes a review of both conventional and emerging techniques, emphasizing their significance in developing state-of-the-art nanoelectronic devices.

2.4.1 Lithography techniques

Lithography is a cornerstone of nanoelectronics fabrication, enabling the patterning of nanostructures on various substrates. Various lithography methods are utilized based on the specific demands for resolution, production speed, and cost-effectiveness in the device fabrication process [104–112].

2.4.1.1 Photolithography

Photolithography stands as the predominant technique in semiconductor fabrication, utilizing ultraviolet (UV) light to project patterns from a photomask onto a photosensitive resist layer applied to the substrate. Photolithography is crucial in the production of ICs, where multiple layers of patterns are precisely aligned to create complex electronic circuits on silicon wafers. Despite its widespread use, photolithography faces limitations in resolution as device features shrink below 10 nm, necessitating the development of alternative methods.

2.4.1.2 Electron beam lithography

Electron beam lithography (EBL) uses a focused electron beam to directly write patterns onto a resist material sensitive to electrons. With its superior resolution compared to photolithography, EBL is ideal for research and the creation of nanoscale devices. It is commonly used in the fabrication of QDs and nanowires, where extremely small features, down to just a few nanometers, are needed. Due to its high resolution, EBL is essential in prototyping and developing novel nanoelectronic devices, although it is typically slower and more expensive than photolithography, limiting its use in large-scale manufacturing.

2.4.1.3 Nanoimprint lithography

Nanoimprint lithography (NIL) is a developing technique that creates patterns by pressing a mold with nanoscale features into a resist layer. Known for its high throughput and costefficiency, NIL can achieve resolutions below 10 nm. This method is widely used in the production of nanophotonic devices like plasmonic sensors and photonic crystals, where accurate nanoscale patterns are essential. Its capacity to replicate complex nanostructures with great precision makes NIL a promising option for large-scale manufacturing of nanoelectronic devices.

2.4.2 Deposition techniques

Deposition methods are used to apply thin material layers onto substrates, forming the essential components of nanoelectronic devices. The selection of a deposition technique is based on the material's properties and the required thickness and uniformity of the film [113–119].

2.4.2.1 Chemical vapor deposition

CVD is a process where gaseous precursors react or decompose on a heated substrate to create a thin film. It is commonly used to deposit high-quality films of semiconductors, insulators, and metals. CVD plays a crucial role in the growth of graphene and CNTs for nanoelectronic applications. By optimizing process parameters, researchers can produce large-area, uniform graphene films or vertically aligned CNT arrays, which are vital for advancing next-generation transistors and interconnects.

2.4.2.2 Atomic layer deposition

Atomic layer deposition (ALD) is a technique that enables the deposition of thin films, one atomic layer at a time through a series of self-limiting surface reactions. It offers exceptional control over film thickness and uniformity, even on intricate 3D structures. ALD is utilized to apply high-k dielectrics in advanced metal-oxide-semiconductor field-effect transistors (MOSFETs), with precise control over the dielectric layer thickness supporting the scaling

Table 2.3 Summary of nanoelectronics and their applications

Characterization method	*Principle*	*Applications in nanoelectronics*	*Ref.*
Scanning Electron Microscopy (SEM)	A focused electron beam to scan the sample's surface. The interaction between the beam and the sample produces secondary electrons, which are detected to create high-resolution images, revealing detailed surface features.	Used to examine the surface morphology and composition of nanomaterials such as CNTs, nanowires, and graphene. Critical for evaluating alignment, diameter, and uniformity in CNT-based transistors	[104, 132]
Transmission Electron Microscopy (TEM)	A high-energy electron beam passes through a very thin sample, creating an image or diffraction pattern from the electrons that scatter as they interact with the sample.	Provides atomic-level images of the internal structure of materials like QDs, revealing size, shape, crystallinity, and defects. Essential for optoelectronic devices such as QD-based LEDs.	[105, 133]
Raman Spectroscopy	Analyzes vibrational modes of molecules by scattering light. Identifies material properties such as defects, doping, and layer count. Extensively used for graphene and CNTs.	Provides information on vibrational modes, layer number, and defects in graphene for optimizing flexible electronics and high-frequency transistors.	[106,134]
X-ray Photoelectron Spectroscopy (XPS)	The material is exposed to X-rays, causing it to release photoelectrons. These emitted photoelectrons are then measured to analyze the material's elemental composition and chemical states.	Used in the study of MOS devices to identify oxidation states, contaminants, and interfacial reactions, critical for improving electrical performance.	[107, 135]
Ultraviolet Visible Spectroscopy (UV-Vis)	It measures how UV or visible light is absorbed by a material to study its electronic transitions and determine the energy gap between its electron levels.	Characterizes semiconductor nanomaterials like cadmium selenide (CdSe) QDs, determining size and optical properties for applications in solar cells, displays, and biological imaging.	[108,136]
Atomic Force Microscopy (AFM)	It scans the surface with a sharp tip to measure the surface shape, creating a map based on the forces between the tip and the surface at the atomic level.	Used to identify flaws in organic thin-film transistors and investigate the mechanical properties and surface roughness of thin films in organic electronics (OTFTs).	[109, 137]

Table 2.3 (*Cont.*)

Characterization method	*Principle*	*Applications in nanoelectronics*	*Ref.*
Scanning Tunneling Microscope (STM)	It measures the current flowing between a metal tip and the sample surface, creating detailed images of the material's electronic properties at the atomic level.	Critical for studying the electronic characteristics of individual atoms and molecules, such as the arrangement of atoms in QDs for potential quantum computing applications.	[110, 138]

of transistors to nanoscale sizes, enhancing device performance and minimizing power consumption [120–122].

2.4.2.3 *Physical vapor deposition*

Physical vapor deposition (PVD) is a technique where material is transferred from a source to a substrate, often through evaporation or sputtering. It is frequently used to deposit metal films that function as contacts or interconnects in nanoelectronic devices. PVD is particularly useful in creating thin metal layers for the production of nanoscale magnetic random-access memory (MRAM) devices. These metal layers form the magnetic tunnel junctions, which are essential for the operation of MRAM, a promising technology for high-speed, non-volatile memory [123–125].

2.4.3 Etching techniques

Etching is a technique used to selectively eliminate material from a substrate to form patterns or structures. Depending on the material and desired resolution, both wet and dry etching methods are used in nanoelectronics [126–131].

2.4.3.1 *Wet etching*

Wet etching involves using liquid chemicals to dissolve specific materials from the substrate. This technique is typically used for materials that can be selectively etched with high precision. Wet etching is used to define the active regions of silicon in the fabrication of metal-oxidesemiconductor (MOS) devices. By carefully controlling the etchant concentration and temperature, precise patterns can be etched into the silicon substrate.

2.4.3.2 *Dry etching*

Dry etching involves the plasma or reactive ion gases to selectively remove material from the substrate. It offers better control over etching profiles and is suitable for creating highly anisotropic (directional) features. Dry etching is used in the fabrication of FinFETs, where narrow, vertical fins of silicon are etched to create the transistor's channel. The ability to precisely etch vertical sidewalls is crucial for maintaining the performance of these nanoscale transistors.

2.4.4 Emerging fabrication techniques

As the demand for ever-smaller and more complex nanoelectronic devices grows, new fabrication techniques are emerging to overcome the limitations of traditional methods.

2.4.4.1 *Self-assembly*

Self-assembly refers to the process in which molecules i.e., nano materials naturally arrange themselves into structures and patterns without the need for external guidance. This technique is being explored for its potential to create complex nanostructures with minimal fabrication steps. Block copolymers, which can self-assemble into nanoscale patterns, are used to develop templates for pattern transfer in semiconductor fabrication. This technique holds the promise of advancing Moore's Law by facilitating the creation of sub-10 nm features for upcoming generations of ICs.

2.4.4.2 *Directed self-assembly*

Directed self-assembly (DSA) merges conventional lithography with self-assembly methods to achieve patterns beyond lithographic resolution limits. Using pre-engineered templates, block copolymers are directed to form specific nanoscale structures. This technique is advancing the production of bit-patterned media for high-density data storage, enabling the precise organization of features essential for compact and efficient data storage solutions.

2.4.4.3 *Nanoimprint lithography*

NIL stands out as a promising approach due to its capability to generate nanoscale features with exceptional precision at a relatively low cost. Its strength lies in fabricating extensive patterns with sub-10 nm accuracy. This technique is instrumental in manufacturing nanophotonics devices and

NEMS, where achieving detailed nanoscale patterns over large surfaces is vital for enhancing device functionality and scalability.

2.5 CHARACTERIZATION METHODS

Characterization techniques play a significant role in assessing the structural and electronic properties of nanoelectronic materials (refer Tables 2.2 and 2.3).

2.6 APPLICATIONS OF NANOELECTRONICS MATERIALS

Nanoelectronics have already begun to have a profound impact on society, driving innovations across multiple industries. From enhancing the performance and efficiency of electronic devices to enabling early disease detection and promoting sustainable energy solutions, the real-world applications of nanoelectronic devices and systems demonstrate the field's potential to continue transforming our world. As research advances, we can expect nanoelectronics to further embed itself into various aspects of daily life, shaping the future of technology and society [139–142].

1. *Transistors in modern computing:* Silicon-based transistors, now scaled down to the nanometer regime, are the fundamental building blocks of modern microprocessors. These nanoscale transistors propel advances in computing power by making it possible to produce computers that are smaller, faster, and more efficient. This miniaturization has been crucial for the development of everything from smartphones to advanced supercomputers, influencing industries such as IT, finance, and healthcare [143].
2. *CNTFETs:* Transistors based on CNTs are being explored as replacements for traditional silicon transistors because of their superior electrical properties. CNTFETs have the potential to enhance device performance and energy efficiency, paving the way for advanced computing and flexible electronics. This could lead to more powerful, energy-efficient devices with a smaller environmental impact [144].
3. *QDs in displays and lighting:* QDs are nanoscale semiconductor particles that exhibit unique optical properties, making them ideal for use in display technologies. QD displays, like those used in high-definition televisions, provide more vibrant colors and greater energy efficiency compared to traditional displays. This technology enhances visual experiences for consumers and reduces energy consumption, contributing to more sustainable electronics [145].
4. *Nanosensors in healthcare:* Nanoscale biosensors capable of detecting biomarkers at very low concentrations are being developed

for early diagnosis of diseases. By facilitating the early detection of diseases like cancer or infectious ones, these nano-biosensors have the potential to completely transform healthcare, resulting in prompt interventions and improved patient outcomes. This development bolsters the movement toward personalized medicine, in which each patient receives a customized course of treatment based on accurate diagnostic data [146].

5. *Energy harvesting devices:* Nanostructured materials like perovskite are used in photovoltaic cells to harvest solar energy more efficiently. These cutting-edge materials are essential for creating solar panels that are more effective than conventional silicon-based solar cells at converting sunlight into electrical power. The widespread adoption of such nano-enabled solar technologies could significantly enhance global renewable energy capacities, contributing to efforts to combat climate change [147,148].
6. *Flexible and wearable electronics:* Nanomaterials like graphene are used in flexible electronic devices, including wearable health monitors and flexible displays. Flexible electronics have numerous applications, from wearable health devices that monitor vital signs to foldable smartphones and displays. These innovations are reshaping how we interact with technology, making devices more adaptable to various environments and uses, and improving the accessibility and personalization of healthcare monitoring [149, 150].
7. *Spintronic devices:* Spintronic devices, which use both the electron's spin and charge, are being created for memory storage that retains data without power, such as non-volatile memory storage (e.g., MRAM). Spintronics offers the possibility of faster, more energy-efficient memory storage solutions with higher data retention capabilities. This technology is poised to enhance data storage systems, leading to faster, more reliable computing devices and potentially revolutionizing data centers and cloud computing infrastructure [151].
8. *Environmental monitoring systems:* Nanosensors designed for environmental monitoring can detect pollutants at the molecular level. These nanosensors facilitate real-time tracking of air and water quality, offering essential information for safeguarding the environment and public health. By detecting contaminants at very low concentrations, they help prevent environmental disasters and ensure safer living conditions, contributing to sustainable development goals [152, 153].

2.7 CHALLENGES AND FUTURE DIRECTIONS

Despite significant progress, nanoelectronic materials face challenges such as the following:

Scalability and manufacturing: Scaling nanoelectronic devices for mass production requires overcoming challenges related to uniformity, yield, and cost. Advances in fabrication techniques and process integration are critical for achieving scalability [154, 155].

Integration with existing technologies: Integrating nanoelectronic components with current semiconductor technologies involves addressing compatibility issues related to materials, fabrication processes, and device architectures [156, 157].

Quantum decoherence: For quantum computing and other quantum-based nanoelectronic devices, maintaining coherence of quantum states over practical timescales is a major challenge [158,159].

Environmental and health considerations: The potential environmental and health impacts of nanomaterials need thorough investigation. Proper guidelines and regulations are necessary to ensure safe handling, use, and disposal [160, 161].

Interdisciplinary collaboration: Advancements in nanoelectronics require collaboration across multiple fields such as physics, chemistry, materials science, and electrical engineering. Interdisciplinary research efforts are essential for addressing complex challenges and driving innovation [162–165].

2.7.1 Recent advances in nanoelectronics

Quantum computing: Recent developments in quantum computing focus on utilizing nanoelectronic devices to generate qubits, the core elements of quantum information. Superconducting, topological, and spin qubits are leading the way in this field of research [166,167].

2D materials beyond graphene: Materials such as MoS_2 and TMDs have shown promise in nanoelectronics. These materials have special electrical, optical, and mechanical qualities that can be adjusted for particular uses [168,169].

Neuromorphic computing: Mimicking the neural architecture of the human brain, neuromorphic systems use nanoelectronic components like memristors to create synapse-like connections. These systems have potential applications in artificial intelligence and machine learning [170,171].

Energy-efficient computing: With the increasing demand for energy-efficient computing, research is focused on developing nanoelectronic devices that consume less power. Tunnel FETs, negative capacitance FETs, and other low-power devices are being studied to see if they can reduce energy consumption in electronic systems [172,173].

Flexible and wearable electronics: The advancement of stretchable and flexible nanoelectronic materials opens up innovative possibilities for applications in wearable technology. These devices can conform to the human body, providing new opportunities in health monitoring and personal electronics [174, 175].

Nanoelectronic-based medical devices: Nanoelectronic devices are revolutionizing medical diagnostics and treatment. Nanobiosensors, implantable devices, and lab-on-a-chip technologies offer unprecedented capabilities in healthcare [176–179].

2.8 CONCLUSION

Nanoelectronics hold great promise for revolutionizing diverse facets of science and society. To conquer present hurdles and reach the maximum potential of nanoelectronic devices, further research and development is required. Nanoelectronic materials are the cornerstone of future technologies, spurring innovation and progress across numerous industries. Through comprehension of the principles of nanoelectronic materials, scientists can open up new avenues for innovation in technology and social effects.

REFERENCES

1. Soleymani, L., and Li, F. "Mechanistic Challenges and Advantages of Biosensor Miniaturization into the Nanoscale." *ACS Sensors*, 2(4), 2017, 458–467.
2. Whitehouse, D.J. "Surface Geometry, Miniaturization and Metrology." *Transactions of the Royal Society A: Mathematical Physical and Engineering Sciences*, 370, 2012, 4042–4065.
3. Jin, Y.M., Wang, Y.U., Khachaturyan, A.G., Li, J., and Viehland, D. "Conformal Miniaturization of Domains with Low Domain-Wall Energy: Monoclinic Ferroelectric States near the Morphotropic Phase Boundaries." *Physical Review Letters*, 91, 2003, 197601.
4. Shekhar, B., and Chien, A. "The Future of Microprocessors." *Communications of the ACM*, 54, 2011, 67–77.
5. Chi-Weon, Y. "The Fundamentals of NAND Flash Memory: Technology for Tomorrow's Fourth Industrial Revolution." *IEEE Solid-State Circuits Magazine*, 14, 2022, 56–65.
6. Daniel Joy, M., and Siva Kumar, K. "Recent Trends and Improvisations in FPGA." *IOSR Journal of Electrical and Electronics Engineering (IOSR-JEEE)*, 12, 2017, 71–77.
7. Deepti, B. "Comparative Study of Various Systems on Chips Embedded in Mobile Devices." *Innovative Systems Design and Engineering*, 4, 2013, 1–9.

8. Balaji Narayan, M. "Advancements in Custom ASIC Development for AI, Edge Computing and IoT Applications." *IJRET*, 11, 2024, 1–8.
9. Amarnath, C.T.K. "Analog and Mixed-Signal Integrated Circuits of 5G and 6G Data Communication." *International Journal for Multidisciplinary Research*, 6, 2024, 1–28.
10. Tushar, S., and Deepak, K. "Data converters for Scalable CMOS Mixed Signal Systems for VLSI." *International Research Journal of Engineering and Technology*, 3, 2016, 2366–2370.
11. Dalvinder, S.G. "Present and Future Nanoelectronics Devices—An Analytical Study." *Research and Development in Material Science*, 8, 2013, 925–930.
12. Niravkumar, D. "Power Electronics Innovations: Improving Efficiency and Sustainability in Energy Systems." *Asia Pacific Journal of Energy and Environment*, 9, 2022, 71–80.
13. Vaishnavi, B., Babaji, G., and Seema, Y. "A Comprehensive Review on Advancements in Wearable Technologies: Revolutionizing Cardiovascular Medicine." *Cureus*, 29, 2024, e61312.
14. He, H. "Overview of Automotive Electronics and Their Development Trends." *IEEE Access*, 8, 2020, 123456–123467.
15. Zhang, L. "Development of a Smart Washing Machine Based on Internet of Things." *Journal of Cleaner Production*, 280, 2021, 124243.
16. Kumar, S. "Smart Refrigerators: A Review of Technologies and Future Directions." *Energy Reports*, 8, 2022, 273–290.
17. Banerjee, A. "Advancements in Smart Air Conditioning Systems: A Review." *Renewable and Sustainable Energy Reviews*, 161, 2023, 112373.
18. Smith, R. "The Role of Electronic Components in Camera Systems." *IEEE Transactions on Consumer Electronics*, 66, 2020, 10–18.
19. Jones, T. "Modern Elevator Systems and Their Electronic Control." *International Journal of Mechanical Engineering and Robotics Research*, 10, 2021, 215–221.
20. Zhao, Y. "Review of Electric Bicycle Technologies." *Journal of Renewable and Sustainable Energy*, 13, 2021, 123–142.
21. Patel, K. "Design and Implementation of a Robotic Arm using Arduino." *Journal of Robotics and Automation*, 7, 2022, 30–39.
22. Johnson, M. "The Role of Electronics in Modern Musical Instruments." *Journal of the Audio Engineering Society*, 69, 2021, 87–95.
23. Smith, J. "Smart Heating Systems: A Review of Technologies and Applications." *Energy Procedia*, 157, 2020, 223–232.
24. Fortes, J.A., Figueiredo, R.J., and Lundstrom, M.S. "Virtual Computing Infrastructures for Nanoelectronics Simulation." *Proceedings of the IEEE*, 93, 2005, 1839–1847.
25. Pargeter, J., and Shea, J.J. "Going Big versus Going Small: Lithic Miniaturization in Hominin Lithic Technology." *Evolutionary Anthropology: Issues*, 28, 2019, 72–85.
26. Fallahpour, M.B., and Zoughi, R. "Antenna Miniaturization Techniques: A Review of Topology- and Material-Based Methods." *IEEE Antennas and Propagation Magazine*, 60, 2018, 38–50.

27. Abhijith, M.N., and Alexander, B. "Analysis of Miniaturization Effects and Channel Selection Strategies for EEG Sensor Networks with Application to Auditory Attention Detection." *IEEE Transactions on Biomedical Engineering*, 67, 2019, 234–244.
28. Mitsudo, K., Kurimoto, Y., Yoshioka, K., and Suga, S. "Miniaturization and Combinatorial Approach in Organic Electrochemistry." *Chemical Reviews*, 118, 2018, 5985–5999.
29. Dick, H.B., Schultz, T., and Gerste, R.D. "Miniaturization in Glaucoma Monitoring and Treatment: A Review of New Technologies That Require a Minimal Surgical Approach." *Ophthalmology and Therapy*, 8, 2019, 19–30.
30. Seng, G.T., and Mays, B.J. "Graphene Carbon Nanostructures for Nanoelectronics. Introduction to the Physics of Nanoelectronics, Introduction to the Physics of Nanoelectronics." *Woodhead Publishing Series in Electronic and Optical Materials*, 2012, 198–242.
31. Laocharoensuk, R., Palaniappan, K., and Smith, A.N. "Flow-Based Solution–Liquid–Solid Nanowire Synthesis." *Nature Nanotechnology*, 8, 2013, 9.
32. Jiayi, H., Han, Y., Min, N., Hai, G., Hongya, Y., Zhongwu, L., and Rong, S. "Soft Magnetic Materials for Power Inductors: State of Art and Future Development." *Materials Today Electronics*, 6, 2023, 100066.
33. Yuqia, R., Yiwen, S., Xionghui, J., Pingfan, G., Zhixuan, C., Yunke, Z., Qi, W., Yu, P., Yanping, L., Yunan, G., and Yu, Y. "Large-Scale Vertically Interconnected Complementary Field-Effect Transistors Based on Thermal Evaporation." *Small*, 2023, e2309953.
34. Morello, A., Pla, J.J., Zwanenburg, F.A., Chan, K.W., Tan, K.Y., Huebl, H., Möttönen, M., Nugroho, C.D., Yang, C., Donkelaar, J.A., Alves, A.D., Jamieson, D.N., Escott, C.C., Hollenberg, L.C., Clark, R.G., and Dzurak, A.S. "Single-Shot Readout of an Electron Spin in Silicon." *Nature*, 467, 2010, 687–691.
35. Naresh, M., and Padmanabham, M. "Design and Analysis of a Heavy Vehicle Chassis for Composite Materials for Optimum Load Conditions." *Journal of Emerging Technologies and Innovative Research*, 5, 2018, 692–696.
36. Kilby, J.S. "Nobel Lecture: Turning Potential into Realities: The Invention of the Integrated Circuit." *Reviews of Modern Physics*, 73, 2000, 639–643.
37. Gordon, E.M. "Cramming More Components onto Integrated Circuits." *Proceedings of the IEEE*, 86, 1998, 82–85.
38. Roukes, M.L. "Nanoelectromechanical Systems." In *Technical Digest MEMS 2001. 14th IEEE International Conference on Micro Electro Mechanical Systems (Cat. No.01CH37090)*, 2001, Transducers Research Foundation, Inc., 387–390.
39. Alivisatos, A.P. "Semiconductor Clusters, Nanocrystals, and Quantum Dots." *Science*, 271,1996, 933–937.
40. Oriols, X. "Quantum Mechanical Effects on Noise Properties of Nanoelectronic Devices: Application to Monte Carlo Simulation." *IEEE Transactions on Electron Devices*, 50, 2003, 1830–1836.

41. Anantram, M.P., Lundstrom, M.S., and Nikonov, D.E. "Modeling of Nanoscale Devices." *Proceedings of the IEEE*, 96(9), 2008, 1511–1550.
42. Supriyo, D. *Lessons from Nanoelectronics: A New Perspective on Transport*. World Scientific Publishing Company First Edition, 2012, 1–18.
43. Blick, R.H. "Quantum Nanoelectronics." In *Nanotechnology*. Springer, 29, 2018, 302001.
44. Dresselhaus, M.S., Dresselhaus, G., and Eklund, P.C. *Science of Fullerenes and Carbon Nanotubes Their Properties and Applications*. Academic Press, San Diego,1996, xviii + 965 pp.
45. Geim, A.K., and Novoselov, K.S. "The Rise of Graphene." *Nature Materials*, 6, 2007, 183–191.
46. Chhowalla, M., Shin, H.S., Eda, G., Li, L.J., Loh, K.P., and Zhang, H. "The Chemistry of Two-Dimensional Layered Transition Metal DichalcogenideNanosheets." *Nature Chemistry*, 5, 2013, 263–275.
47. Wolf, S.A., Awschalom, D.D., Buhrman, R.A., Daughton, J.M., von Molnár, S., Roukes, M.L., and Parkin, S.S.P. "Spintronics: A Spin-Based Electronics Vision for the Future." *Science*, 294, 2001, 1488–1495.
48. Michalet, X., Pinaud, F.F., Bentolila, L.A., Tsay, J.M., Doose, S., Li, J.J., and Weiss, S. "Quantum Dots for Live Cells, In Vivo Imaging, and Diagno stics."*Science*,307, 2005, 538–544.
49. Kastner, M.A. "The Single-Electron Transistor." *Reviews of Modern Physics*, 64, 1993, 849–858.
50. Foster, A.S., and Shluger, A.L. "Scanning Tunneling Microscopy: Atomic Scale Imaging and Manipulation." *Journal of Physics: Condensed Matter*, 13, 2001, R973–R1000.
51. David, B.W., and Carter, C.B. "Transmission Electron Microscopy." In*A Textbook for Materials Science*. Springer, LXII, 2009, 775.
52. Buhrman, R.A., and Ralph, D.C. "Spintronics: A New Way to Store Data." *Physics Today*, 56, 2003, 58–64.
53. Novoselov, K.S.,Geim, A.K., Morozov, S.V., Jiang, D., Zhang, Y., Dubonos, S.V., and Firsov, A.A. "Electric Field Effect in Atomically Thin Carbon Films." *Science*, 306, 2004, 666–669.
54. Zhang, H., Wang, C., Lin, J., Tian, X., and Shen, X. "Tunneling Field-Effect Transistors: Physics, Materials and Devices." *Micromachines*, 11, 2020, 731.
55. Lihong, S., Zhou, Y., Xitong, W., Ziao, Z., Bo, W., Gary, H., Ninghui, C., Yongyong, S., Zhibo, M., Haoxu, W., Yucheng, L., Junping, Z., Shuanhu, W., Yuefei, L., Fengxia, Y., Jixin, Z., Fei, G., Wei, H., and Shengzhong, L. "Flexible Diodes/Transistors Based on Tunable p–n-Type Semiconductivity in Graphene/Mn–Co–Ni–O Nanocomposites." *Research (Washington, DC)*, October 13, 2021, 9802795.
56. Binnig, G., and Rohrer, H. "Scanning Tunneling Microscopy." *Surface Science*, 126, 1983, 236–244.
57. Brinkman, W. F., Dynes, R. C. and Rowell J. M. " Tunneling Conductance of Asymmetrical Barriers " *Journal of Applied Physics*, 41, 1970, 1915–1921.
58. Baker, S. "Wave-Particle Duality: A Quantum Perspective in Nanoelectronics." *Journal of Applied Physics*, 125, 2019, 180901.

59. Nagaev, K.E. "Quantum Transport and Wave-Particle Duality in Nanoscale Devices." *Physics Uspekhi*, 172, 2002, 231–244.
60. Ghosh, A., and Das, A. "Exploring Wave-Particle Duality in Quantum Dots: Implications for Nanoelectronics." *Nature Nanotechnology*, 15, 2020, 121–129.
61. Klein, D.J., and Hsu, Y.T. "Quantum Mechanics in Nanoelectronics: Wave-Particle Duality and Its Applications." *Applied Physics Reviews*, 5, 2018, 041106.
62. Garg, A. "Heisenberg's Uncertainty Principle and Its Implications in Quantum Tunneling." *Journal of Physics: Conference Series*, 1060, 2018, 012025.
63. Kane, B.E. "A Silicon-Based Quantum Computer." *Nature*, 442, 2007, 696–699.
64. Liu, R., and Xu, H. "Quantum Tunneling in Nanoscale Devices: The Role of the Uncertainty Principle." *Applied Physics Letters*, 114, 2019, 103101.
65. Yuli, V.N.,and Yaroslav, M.B. *Quantum Transport: Introduction to Nanoscience*. Cambridge University Press,2009, 581.
66. Koch, S.W., and Vardeny, Z.V. "Fundamentals of Scanning Tunneling Microscopy." *Annual Review of Physical Chemistry*, 67, 2016, 373–397.
67. Binnig, G., and Rohrer, H. "Scanning Tunneling Microscopy." *Physical Review Letters*, 56, 1986, 930–933.
68. Zhou, C., and Ma, Z. "Tunnel Diodes: Principles, Applications, and Emerging Trends." *IEEE Transactions on Electron Devices*, 64, 2017, 1283–1294.
69. Klein, D.J., and Tikhodeev, S. "Quantum Conductance in One-Dimensional Nano-Devices." *Nature Reviews Physics*, 1, 2018, 76–90.
70. Buttiker, M. "Quantum Transport: The Non-Equilibrium Green's Function Method." *Physical Review Letters*, 65, 1990, 290–293.
71. Kane, B.E. "A Silicon-Based Quantum Computer." *Nature*, 393, 1998, 133–137.
72. Loss, D., and Di Vincenzo, D.P. "Quantum Computation with Quantum Dots." *Physical Review A*, 57, 1998, 120–126.
73. Michael, A.N., and Isaac, L.C. *Quantum Computation and Quantum Information*. Cambridge University Press, 2010, 10th Edition, 1–710.
74. Charles,H.B., and Gilles, B. "Quantum Cryptography: Public Key Distribution and Coin Tossing." *Theoretical Computer Science*, 560, Part 1, 4 December 2014, 7–11.
75. Gisin, N., Ribordy, G., Tittel, W., and Zbinden, H. "Quantum Cryptography." *Reviews of Modern Physics*, 74, 2002, 145–195.
76. Novoselov, K.S., Geim, A.K., Morozov, S.V., Jiang, D., Zhang, Y., Dubonos, S.V., and Grigorieva, I.V. "Electric Field Effect in Atomically Thin Carbon Films." *Science*, 306, 2004, 666–669.
77. Iijima, S. "Helical Microtubules of Graphitic Carbon." *Nature*, 354, 1991, 56–58.
78. Cui, Y., and Lieber, C.M. "Functional Nanoscale Electronic Devices Assembled Using Silicon Nanowire Building Blocks." *Science*, 291, 2001, 851–853.

79. Murray, C.B., Kagan, C.R., and Bawendi, M.G. "Synthesis and Characterization of Monodisperse Nanocrystals and Close-Packed Nanocrystal Assemblies." *Annual Review of Materials Research*, 30, 2000, 545–610.
80. Wang, Q.H. "Electronics and Optoelectronics of Two-Dimensional Transition Metal Dichalcogenides." *Nature Nanotechnology*, 7, 2012, 699–712.
81. Dean, C.R. "Boron Nitride Substrates for High-Quality Graphene Electronics." *Nature Nanotechnology*, 5, 2010, 722–726.
82. Andrews, R. "Multiwall Carbon Nanotubes: Synthesis and Application." *Accounts of Chemical Research*, 35, 2002, 1008–1017.
83. Furukawa, H. "The Chemistry and Applications of Metal-Organic Frameworks." *Science*, 341, 2013, 1230444.
84. Hussain, F. "Polymer-Matrix Nanocomposites, Processing, Manufacturing, and Application: An Overview." *Journal of Composite Materials*, 40, 2006, 1511–1575.
85. Avouris, P., Chen, Z., and Perebeinos, V. "Carbon-Based Electronics." *Nature Nanotechnology*,2, 2007, 605–615.
86. Thess, A., Lee, R., and Nikolaev, P. "Crystalline Ropes of Metallic Carbon Nanotubes." *Science*, 273, 1996, 483–487.
87. De Volder, M., Tawfick, S., Baughman, R.H., and Hart, A.J. "Carbon Nanotubes: Present and Future Commercial Applications." *Science*, 339, 2013, 535–539.
88. Nikolaev, P.,Bronikowski, M.J., and Bradley, R.K. "Gas-Phase Catalytic Growth of Single-Walled Carbon Nanotubes from Carbon Monoxide." *Chemical Physics Letters*, 313, 1999, 91–97
89. Cheng, H.M., Li, F., and Su, G. "Large-Scale and Low-Cost Synthesis of Single-Walled Carbon Nanotubes by the Catalytic Decomposition of C_2H_2 over Fe–Mo/MgO Catalyst." *Journal of Nanoscience and Nanotechnology*, 3, 2003, 285–292.
90. Li, Y., Kinloch, I.A., and Windle, A.H. "Direct Spinning of Carbon Nanotube Fibers from Chemical Vapor Deposition Synthesis." *Nano Letters*, 4, 2004, 317–321.
91. Yu, A. "Large-Scale Production of Carbon Nanotubes by Electrolysis." *Chemical Physics Letters*, 323, 2000, 329–334.
92. Chen, B. "The Production of Carbon Nanotubes by Ball Milling." *Carbon*, 37, 1999, 1865–1873.
93. Vander Wal, R.L., and Ticich, T.M. "Flame and Furnace Synthesis of Single-Walled and Multi-Walled Carbon Nanotubes and Nanofibers." *Carbon*, 39, 2001, 2277–2289.
94. Ajayan, P.M., and Iijima, S. "Capillarity-Induced Filling of Carbon Nanotubes." *Nature*, 375, 1995, 564–567.
95. Mittal, V., Vaish, R., and Tanaka, K. "Graphene-Based Polymer Nanocomposites: A Review." *Journal of Polymer Science, Part B: Polymer Physics*, 53, 2015, 1031–1048.
96. Castro Neto, A.H., Guinea, F., Peres, N.M.R., Novoselov, K.S., and Geim, A.K. "The Electronic Properties of Graphene." *Reviews of Modern Physics*, 81, 2009, 109–162.

97. Cui, Y., and Lieber, C.M. "Functional Nanoscale Electronic Devices Assembled Using Silicon Nanowire Building Blocks." *Science*, 291, 2001, 851–853.
98. Murray, C.B., Kagan, C.R., and Bawendi, M.G. "Synthesis and Characterization of Monodisperse Nanocrystals and Close-Packed Nanocrystal Assemblies." *Annual Review of Materials Research*, 30, 2000, 545–610.
99. Wang, Q.H., Kalantar-Zadeh, K., Kis, A., Coleman, J.N., and Strano, M.S. "Electronics and Optoelectronics of Two-Dimensional Transition Metal Dichalcogenides." *Nature Nanotechnology*, 7, 2012, 699–712.
100. Dean, C.R., Young, A.F., and Cadden-Zimansky, K. "Boron Nitride Substrates for High-Quality Graphene Electronics." *Nature Nanotechnology*, 5, 2010, 722–726.
101. Furukawa, H., Gandara, F., and Zhang, Y.B. "The Chemistry and Applications of Metal-Organic Frameworks." *Science*, 341, 2013, 1230444.
102. Andrews, R., and Jacques, D.A. "Multiwall Carbon Nanotubes: Synthesis and Application." *Accounts of Chemical Research*, 35, 2002,1008–1017.
103. Hussain, F., Hojjati M., Okamoto M. and Gorga R.E."Polymer-Matrix Nanocomposites, Processing, Manufacturing, and Application: An Overview." *Journal of Composite Materials*, 40, 2006, 1511–1575.
104. Wong, H.-S.P., and Salahuddin, S. "Memory Leads the Way to Better Computing." *Nature Nanotechnology*, 10, 2015, 191–194.
105. Moore, G.E. "Cramming More Components onto Integrated Circuits." *Electronics*, 38, 1965, 114–117.
106. Chen, X., Xu, X., Li, T., and Guo, D. "Extreme Ultraviolet Lithography: A Review." *Micromachines*, 12, 2021,414–421.
107. Van den Heuvel, D.B., and Klein, C.A. "Advancements and Alternatives in Semiconductor Lithography Beyond Photolithography." *Journal of Micro/Nanolithography, MEMS, and MOEMS*, 20, 2021, 011015.
108. Guo, L.J. "Nanoimprint Lithography: Methods and Material Requirements." *Advanced Materials*, 19, 2007, 495–513.
109. Chou, S.Y., Krauss, P.R., and Renstrom, P.J. "Nanoimprint Lithography." *Journal of Vacuum Science& Technology B: Microelectronics and Nanometer Structures Processing, Measurement and Phenomena,* 14, 1996, 4129–4133.
110. Hua, F., Sun, Y., Gaur, A., Meitl, M.A., Bilhaut, L., Rotkina, L., and Rogers, J.A. "PolymerImprintLithography with Molecular-Scale Resolution." *Nano Letters*, 4, 2004, 2467–2471.
111. Resnick, D.J., Sreenivasan, S.V., and Willson, C.G. "Step and Flash Imprint Lithography for Sub-100-nm Patterning Applications." *Proceedings of the IEEE*, 93, 2005, 1500–1510.
112. Zhu, X., Li, X., and Zhang, D. "Application of Nanoimprint Lithography in the Fabrication of Plasmonic Nanostructures for Sensors." *Journal of Nanoscience and Nanotechnology*, 16,2006, 123–132.
113. Deckman, H.W., and Dagan, Z. "Chemical Vapor Deposition of Thin Films: The State of the Art for Large Area Applications." *Applied Physics Reviews*, 1, 2014, 041302.

114. Li, X., Cai, W., An, J., Kim, S., Nah, J., Yang, D., and Ruoff, R.S. "Large-Area Synthesis of High-Quality and Uniform Graphene Films on Copper Foils by Chemical Vapor Deposition." *Science*, 324, 2009, 1312–1314.
115. Chhowalla, M., Jena, D., and Zhang, H. "Two-Dimensional Semiconductors for Transistors." *Nature Reviews Materials*, 1, 2016, 16052.
116. Kang, S.J., Kocabas, C., Kim, H.S., Cao, Q., Meitl, M.A., Khang, D.Y., and Rogers, J.A. "Printed Multilayer Superstructures of Aligned Single-Walled Carbon Nanotubes for Electronic Applications." *Nature Nanotechnology*, 2, 2007, 230–236.
117. Yoon, Y., and Tiwari, A. "Chemical Vapor Deposition of Graphene: Challenges and Prospects." *Progress in Materials Science*, 83, 2015, 77–126.
118. Tesch, J., and Albrecht, O. "Carbon Nanotubes and Graphene by Chemical Vapor Deposition for Electrochemical Applications." *ChemElectroChem*, 6, 2019, 2776–2800.
119. Dresselhaus, M.S., and Dai, H. "Carbon Nanotubes: Continued Innovations and Challenges." *Journal of the American Chemical Society*, 125, 2003, 1140–1147.
120. George, S.M. "Atomic Layer Deposition: An Overview." *Chemical Reviews*, 110, 2010, 111–131.
121. Kukli, K., Aarik, J., Aidla, A., Uustare, T., and Ritala, M. "Atomic Layer Deposition of Hafnium Oxide Films from Hafnium Tetrakis(Dimethylamide) and Water." *Thin Solid Films*, 408, 2001, 57–62.
122. Leskela, M., and Ritala, M. "Atomic Layer Deposition (ALD): From Precursors to Thin Film Structures." *Thin Solid Films*, 409, 2002, 138–146.
123. Mattox, D.M. *Handbook of Physical Vapor Deposition (PVD) Processing: Film Formation, Adhesion, Surface Preparation, and Contamination Control*. Elsevier Science,2010.
124. Wolf, S.A., and Lu, J.J. "Spintronics and Spin-Transport Magnetic Tunnel Junctions." *Proceedings of the IEEE*, 96, 2008, 1951–1971.
125. Engel, B.N., Akerman, J., Butcher, T., Dave, R.W., DeHerrera, M., Durlam, M., and Slaughter, J.M. "A 4-Mb Toggle MRAM Based on a Novel Bit and Switching Method." *IEEE Transactions on Magnetics*, 41, 2005, 132–136.
126. Choi, C.K., Lee, Y.H., and Kang, B.S. "Wet Chemical Etching of Silicon and Its Compounds for MEMS Applications." *Journal of Micromechanics and Microengineering*, 3, 1993, 76–84.
127. Wolf, S., and Tauber, R.N. *Silicon Processing for the VLSI Era: Volume 1—Process Technology*. Lattice Press,1986.
128. Seidel, H., Csepregi, L., Heuberger, A., and Baumgartel, H. "Anisotropic Etching of Crystalline Silicon in Alkaline Solutions: Orientation Dependence and Behavior of Passivation Layers." *Journal of the Electrochemical Society*, 137, 1990, 3612–3626.
129. Oehrlein, G.S., and Rieger, D. "Fundamentals and Applications of Reactive Ion Etching in Integrated Circuit Processing." *Journal of Vacuum Science & Technology B*, 5, 1987, 1381–1388.
130. Chapman, B.N. *Glow Discharge Processes: Sputtering and Plasma Etching*. Wiley-Interscience,1980.

131. Matsui, S., and Ochiai, Y. "Focused Ion Beam Applications to Nanodevice Fabrication." *Journal of Vacuum Science & Technology B*, 26, 2008, 2549–2558.
132. Giovanni, B., Elia, A.G., Stéphane, B., Tim, W.,Dalia, D., and Mark, W.T. "Supramolecular Reinforcement of Polymer-Nanoparticle Hydrogels for Modular Materials Design." *Advanced Materials*, 2022, 34, e2106941.
133. Anila, F., Tabeer, K., and Azra, Y. "Microbial Synthesis of Nanoparticles and Their Potential Applications in Biomedicine." *Journal of Applied Biomedicine*,15, 2017, 241–248.
134. Antonio, H.C.N., and Francisco, G.F. "The Electronic Properties of Graphene." *Reviews of Modern Physics*,81, 2009, 109.
135. Jin-Ho,L.,Soo-Jeong,P.,and Jeong-Woo,C. "Electrical Property of Graphene and Its Application to Electrochemical Biosensing."*Nanomaterials (Basel)*, 20, 2019, 297.
136. Tsuneya, A. "The Electronic Properties of Graphene and Carbon Nanotubes." *NPG Asia Materials,* 1, 2009, 17–21.
137. Yang, R., Xu, J., Xu, L., Sun, X., Chen, Q., Zhao, Y., Peng, R., and Liu, Z. "Cancer Cell Membrane-Coated Adjuvant Nanoparticles with Mannose Modification for Effective Anticancer Vaccination." *ACS Nano*, 26, 2018, 5121–5129.
138. Rong, Y., Jun, X., Ligeng, X., Xiaoqi, S., Qian, C., Yuhuan, Z., Rui, P., and Zhuang, L. "Cancer Cell Membrane-Coated Adjuvant Nanoparticles with Mannose Modification for Effective Anticancer Vaccination." *ACS Nano*, 2018, 5121–5129.
139. Fiori, G., and Iannaccone, G. "Nanoelectronics: Materials, Devices, Applications." *Microelectronics Journal*, 38, 2007, 311–319.
140. Haichao, L., Ranran, J., Hongbo, C., Xiaolong, T., Changlong, S., Jing, Z., Zhaogang, Y., Jingyao, S. and Chuansheng, W., "Application of Biodegradable and Biocompatible Nanocomposites in Electronics: Current Status and Future Directions" *Nanomaterials,* 9, 2019, 950.
141. Blais, A. "Circuit Quantum Electrodynamics." *Reviews of Modern Physics*, 93, 2021, 025003.
142. Peng-jie, Z., Ya-ni, Y., Yi, Z., Zhi-jie, Q., Yong-feng, L. and Cheng-Meng, C. "Thermal Properties of Graphene and Graphene-Based Nanocomposites: A Review" *ACS Applied Nano Materials,* 7, 2024, 8445–8463.
143. Liu, Y. "Nonvolatile Memory Devices Based on Conductive Polymer Nanocomposites." *Advanced Materials*, 25, 2013, 5497–5502.
144. Dong, Y.Y., and Tae, W.K. "Nonvolatile Memory Devices Based on Au/Graphene Oxide Nanocomposites with Bilateral Multilevel Characteristics." *Carbon*, 88, 2015, 26–32.
145. Mee Rahn, K. and Dongling, M., "Quantum-Dot-Based Solar Cells: Recent Advances, Strategies, and Challenges" *Journal of Physical Chemistry Letters,* 6, 2015, 85–99.
146. Wang, X., "A New Family of Nanowire Sensors for Monitoring Environmental Contaminants." *Environmental Science & Technology*, 46, 2012, 7631–7638.

147. Tiwari, R.K., Mishra, A.K., and Singh, R.P. "Nanostructured Energy Harvesting Devices: Design, Fabrication, and Applications." *Journal of Nanotechnology and Nanomaterials*, 8, 2017, 55–70.
148. Wang, Z.L., and Wu, W. "Nanostructures of Zinc Oxide for Energy Harvesting and Piezotronics."*Advanced Materials*, 24, 2012, 4632–4646.
149. Park, S., "Flexible Graphene-Based Field-Effect Transistors for Low-Noise Applications." *Advanced Functional Materials*, 23, 2013, 2552–2558.
150. Tanmoy, D., Bhupendra K. S., Ajit K. K. and Jong-Hyun, A. "Graphene-based Flexible and Wearable Electronics" *Journal of Semiconductors, 39*, 2018, 011007.
151. Simon, P., and Gogotsi, Y. "Materials for Electrochemical Capacitors." *Nature Materials*, 7, 2008, 845–854.
152. Wang, S., Chen, Z., and Xu, Y. "Nanosensors for Environmental Monitoring: Applications and Advancements." *Environmental Science: Nano*, 7, 2020, 927–948.
153. Liu, Y., Huang, C., Zhang, H., and Zhang, S. "Recent Advancements in Nanosensorsfor Environmental Monitoring: A Review." *Sensors and Actuators B: Chemical*, 283, 2019, 249–261.
154. Chau, R., Doyle, B., Datta, S., Kavalieros, J., and Zhang, K. "Integrated Nanoelectronics for the Future: High-Performance at Reduced Power." *Nature Materials*, 6, 2007, 810–812.
155. Ionescu, A.M., and Riel, H. "Tunnel Field-Effect Transistors as Energy-Efficient Electronic Switches." *Nature*, 479, 2011, 329–337.
156. Tanskanen, J.T., and Lehtonen, M. "Integration of Nanoelectronic Components into Flexible Substrates for Wearable Technologies." *Nano Research*, 11, 2018, 3028–3040.
157. Knezevic, I., and Chen, Z. "Nanoelectronic Device Integration for Quantum Computing Applications." *IEEE Transactions on Nanotechnology*, 18, 2019, 515–524.
158. Zurek, W.H. "Decoherence, Einselection, and the Quantum Origins of the Classical." *Reviews of Modern Physics*, 75, 2003, 715–775.
159. Schlosshauer, M. "Quantum Decoherence: An Introduction and Selected Applications." *Physics Reports*, 831, 2019, 1–57.
160. Colvin, V.L. "The Potential Environmental Impact of Engineered Nanomaterials." *Nature Biotechnology*, 21, 2003, 1166–1170.
161. Donaldson, K., Stone, V., Tran, C.L., Kreyling, W., and Borm, P.J."Nanotoxicology: A New Frontier in Nanotechnology Safety." *Occupational and Environmental Medicine*, 61, 2004, 727–728.
162. Huang, Y., Duan, X., Cui, Y., and Lieber, C.M. "Gallium Nitride Nanowire-Based Nanoelectronics." *Science*, 291, 2001, 630–633.
163. Kumar, S., and Ramanathan, S. "Neuromorphic Devices for Nanoelectronics: A Collaborative Approach." *Nature Nanotechnology*, 12, 2017, 507–514.
164. Schulz, M., Decker, B., and Fuchs, F. "Interdisciplinary Efforts in the Development of Flexible Nanoelectronics." *Advanced Materials*, 27, 2015, 1109–1115.

165. Markovic, I., Glass, G., and Singh, D. "Nanomedicine Meets Nanoelectronics: Collaborative Innovations in Biomedical Devices." *Journal of Nanomedicine & Nanotechnology*, 11, 2020, 220–234.
166. Arute, F., Arya, K., Babbush, R., Bacon, D., Bardin, J.C., and Barends, R. "Quantum Supremacy Using a Programmable Superconducting Processor." *Nature*, 574, 2019, 505–510.
167. Preskill, J. "Quantum Computing in the NISQ Era and Beyond." *Quantum*, 2, 2018, 1–19.
168. Manzeli, S., Ovchinnikov, D., Pasquier, D., Yazyev, O.V., and Kis, A. "2D Transition Metal Dichalcogenides." *Nature Reviews Materials*, 2, 2017, 17033.
169. Xu, M., Liang, T., Shi, M., and Chen, H. "Graphene-Like 2D Materials: Phosphorene, Silicene, and Germanene." *Chemical Reviews*, 113, 2013, 3766–3798.
170. Furber, S. "Large-Scale Neuromorphic Computing Systems." *Frontiers in Neuroscience*, 10, 2016, 28–35.
171. Merolla, P.A., Arthur, J.V., and Alvarez-Icaza, R. "A Digital Neurosynaptic Core Based on the 45 nm CMOS Process." *Science*, 345, 2014, 668–673.
172. Georgescu, A., and Cottet, A. "Energy-Efficient Computing: The Role of Quantum Computing in Future Technologies." *Nature Communications*, 10, 2019, 1144.
173. Hsu, C.H., Huang, K., and Chen, Y. "Recent Developments in Energy-Efficient Computing Technologies." *IEEE Transactions on Emerging Topics in Computing*, 8, 2020, 267–280.
174. Rogers, J.A., and Huang, Y. "Flexible and Stretchable Electronics: A Review of Recent Advances and Future Perspectives." *Advanced Materials*, 31, 2019, 1801268.
175. Kim, J., Lee, H., and Kim, S. "Wearable Electronics for Health Monitoring: Recent Advances and Future Directions." *Journal of Applied Polymer Science*, 137, 2020, 48871.
176. Liu, Y., and Huang, H. "Nanoelectronic Devices for Biomedical Applications: A Review." *Nano Today*, 35, 2020, 100938.
177. Cheng, M., and Zhang, H. "Nanoelectronics in Healthcare: Advances in Diagnostics and Therapeutics". *Advanced Healthcare Materials*, 10, 2021, 2001641.
178. Fischer, J.E., and Lichtenstein, A. "Towards Smart Medical Devices: Recent Innovations in Nanoelectronics." *Nature Reviews Materials*, 4, 2019, 551–570.
179. Wang, J., and Lu, Y. "Nanomaterial-Based Devices for Drug Delivery and Imaging Applications." *Small*, 18, 2022, 2107484.

Chapter 3

Physical and electrical properties of nanomaterials

D.M. Tejashwini, Ramachandra Naik, H.P. Nagaswarupa, and Asmita Deshpande

3.1 INTRODUCTION TO NANOMATERIALS

Nanomaterials are substances that have nanoscale dimensions, usually between 1 and 100 nm. Materials with these dimensions display different physical, chemical, electrical, and optical characteristics from their bulk equivalents [1]. Nanomaterials' exceptional qualities make them essential to the advancement of technology, particularly in nanoelectronics, energy storage, medicinal applications, and environmental science. Nanoelectronics, in particular, benefits greatly from nanomaterials because of their ability to develop quicker, more efficient, and smaller electrical devices [2].

Nanomaterials take several forms, such as small particles, nanotubes, nanowires, and nanofilms. These substances can be inorganic, organic, or hybrid systems, which combine two or more components to improve their characteristics. Carbon-based nanostructures (e.g., graphene, fullerenes, and carbon nanotubes (CNTs)), metal oxide nanoparticles (such as ZnO, TiO_2, and Fe_3O_4), and quantum dots (QDs) in particular are among the most researched nanomaterials [3]. Nanomaterials may have exceptional features due to their high surface-area-to-volume ratio, quantum phenomena, and changed electrical behavior at the nanoscale. This increased area of surface improves chemical reactivity, catalytic activity, and energy efficiency. In the meantime, quantum-confined effects frequently provide size-dependent photonic and electrical features that are not found in bulk materials. Minimization of electrical equipment is a driving factor [4].

Carbon nanomaterials, such as graphene and CNTs, have excellent prospects for nanoelectronics. Graphene, a single layer made up of carbon atoms organized in a hexagonal structure, is electrically highly conductive, mechanically strong, and flexible. Its strong mobility of electrons and ballistic transportation capabilities make it an excellent choice for transistors, sensors, and energy devices. CNTs, on the other hand, display both metallic and semiconducting qualities depending on their structure, thus making them suitable for use in transistors, interconnects, and other electronic devices [5]. Nanowires, yet another kind of nanomaterial, have outstanding

DOI: 10.1201/9781003512899-3

electrical and optical characteristics and are utilized in field-effect transistors (FETs), photodetectors, and nanoscale sensors. Semiconductor nanowires, especially those made from silicon, zinc oxide, or gallium nitride, have emerged as viable alternatives to standard silicon devices, with the ability to enable the development of adaptable, translucent, and low-power devices electronic devices [6]. Metallic nanoparticles, including gold and silver nanoparticles, have distinct electrical characteristics that make them useful in molecular electronics, plasmon resonance, and catalysts for nanodevices. QDs, semiconductor nanoparticles with quantum confinement effects, are important in nanoelectronics because of their size-tunable photonic and electrical characteristics. They have potential uses in displays, solar power, and quantum computers [7].

Nanomaterials have unique features that cannot be observed in their bulk equivalents, owing to their small size, quantum effects, and high surface-to-volume ratio. When materials are reduced to the nanoscale, the behavior of atoms and electrons changes dramatically, resulting in a variety of distinct physical, chemical, electrical, and visual characteristics. Nanomaterials may be critical in the advancement of technology including nanotechnology, sensors, storage of energy, catalytic processes, and biomedicine due to their unique features. Nanoscale materials have a substantially greater surface-to-volume ratio than bulk materials. The majority of atoms in ordinary substances are found on the inside surface, whereas nanomaterials have a substantial fraction of atoms on the surface. The increased surface area considerably improves the material's chemical responsiveness and catalytic effectiveness [8]. For example, nanoparticles of gold or platinum, which are rather harmless in bulk, have extraordinary catalytic capabilities at the nanoscale. This feature is extremely useful for catalysis, sensors, and energy systems. The higher surface area has an impact on mechanical qualities as well. Nanomaterials offer higher strength and hardness because the surface atoms are more firmly linked and less susceptible to distortion. For example, nanoparticles of materials such as ceramics or metals demonstrate higher durability [9].

The encapsulation of electrons in a small space at the nanoscale modifies a material's electrical characteristics. Quantum entanglement occurs when the nanomaterial's size approaches the electron's de Broglie wavelength, resulting in discrete energy levels rather than the continuous energy bands found in bulk materials. Consequently, nanomaterials have size-dependent optical and electrical characteristics. For example, semiconductor QDs show this effect strongly [10]. Their optical characteristics, including both absorption and emission wavelengths, may be easily adjusted by modifying the QDs' size. Smaller QDs produce light at a shorter wavelength (blue), but bigger ones emit at deeper wavelengths (red). This phenomenon is used in applications including QD displays, medical imaging, and solar power cells [11].

3.2 PHYSICAL PROPERTIES OF NANOMATERIALS

Nanomaterials' physical characteristics are strongly impacted by their size and form, separating their properties from their counterparts in bulk as shown in Figure 3.1. When materials are lowered to the nanoscale (usually 1–100 nm), their ratio of surface to volume dramatically increases. This increases surface responsiveness, making nanomaterials more biologically active, which is particularly useful for purposes such as catalysis and sensing [12].

3.2.1 Size dependence

The tiny scale of nanoparticles causes quantum-confined effects, especially in semiconductors and metals. The confinement of quantum particles restricts electron mobility, producing distinct energy levels as well as size-dependent optic and electrical features. As is the case in semiconductor QDs, the bandgap increases as the size of particles decreases, thereby allowing optical features like light emission color to be tuned. Mechanical qualities, including hardness and strength, improve as the size of nanoparticles decreases, owing to fewer structural flaws and grain boundaries in materials such as nanocrystalline metals. Electrical features are similarly affected. The decreased dimensions of nanowires and thin films affect the movement of electrons, thus increasing resistance and providing ballistic propagation in some circumstances [13].

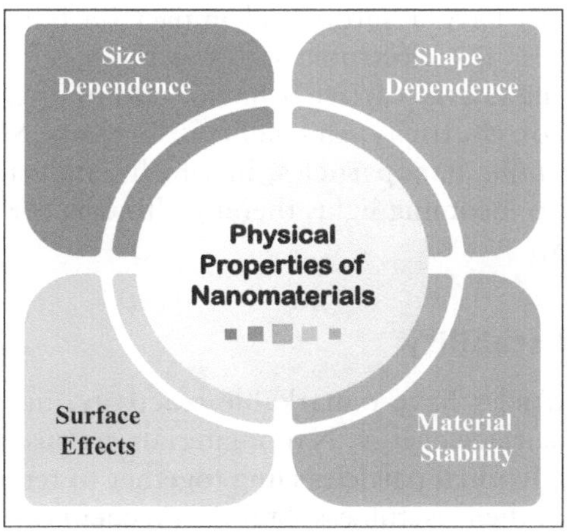

Figure 3.1 Different types of physical properties of nanomaterials.

3.2.2 Shape dependence

Nanomaterials' physical behavior is influenced by their form, which might be spherical, rod-like, or planar. In particular, nanowires and nanorods frequently have anisotropic characteristics, which means that their behavior varies along various axes. Nanowires are tiny in electronics and provide direct electron transport channels, thereby enhancing conductivity throughout their length. Shape impacts optical qualities as well. Localized surface plasmon resonance (LSPR) in metallic nanoparticles is shape-dependent, allowing for variable light absorption in silver or gold nanorods as opposed to spheres. The shape also influences the exposure of catalytic sites, with specific crystal facets of nanocubes or plates giving more active sites, thus improving catalytic efficiency. Surface effects play an important role in defining the physical and chemical properties of materials at the nanoscale. Nanomaterials have a higher surface-to-volume ratio than bulk materials, therefore a considerable number of their atoms are positioned at or near the outermost layer, resulting in different characteristics [14]. These surface atoms have unsaturated bonds, which increases their reactivity and influences phenomena like energy from the surface, enzyme activity, and materials stability.

3.2.3 Surface effects

Nanomaterials have increased surface reactivity because there are fewer nearby atoms to connect with, leading to higher surface energy. This increased energy frequently causes nanoparticles to engage in chemical processes, making them extremely useful in applications involving catalysis. Metal nanoparticles, which include platinum or gold, can catalyze processes more effectively because they have plenty of active interface sites. Surface effects have a substantial impact on both optical and electrical characteristics. The electronic states at the surface of nanomaterials are altered, affecting charge carrier behavior and improving characteristics like conductivity of electricity and light absorption. Surface effects also cause LSPR in metallic nanoparticles, in which conduction electrons fluctuate in response to incoming light, thereby allowing for adjustable optical characteristics [15].

3.2.4 Material stability

Although nanoparticles have remarkable reactivity, instability may result from their higher surface energy. Nanomaterials are susceptible to accumulation, in which individual particles cling together to reduce surface energy, potentially limiting their usefulness. Stabilizing chemicals, surface coatings, or changes are frequently necessary to avoid aggregation and preserve desired qualities. Furthermore, temperature stability can be lowered on the nanoscale. Nanomaterials' high surface energy allows them to undergo transitions in phase at lower temperatures. This presents issues for preserving structural integrity in applications that require high temperatures. Mechanical qualities, which include strength, elasticity, and hardness, are essential characteristics of materials that influence their applicability for diverse uses.

Strength is a material's capacity to bear an applied force without failing. It is often assessed in terms of tensile strength (resistance to separation) or compressive strength (resistance to compression). Bridges, buildings, and machines all require high levels of strength. The highest stress that a material can withstand without shattering is known as its ultimate strength. Steel and titanium are examples of common strong materials used in high-stress settings [16].

Mechanical qualities, which include strength, elasticity, and hardness, are essential characteristics of materials that influence their applicability for diverse uses. Strength is a material's capacity to bear an applied force without failing. Tensile strength (resistance to being ripped apart) and compressive strength (resistance to being crushed) are two regularly used measurements. The bridges, buildings, and machines all require high levels of strength. The highest stress that a material can withstand without shattering is known as its ultimate strength. Steel and titanium are examples of common durable substances used in high-stress settings. A material's hardness is defined as its resistance to deformation, indentation, or scratching [17]. Hardness testing determines how effectively an item will function in environments wherein surface endurance is critical. Because of their resistance to wear, harder materials such as ceramics, diamonds, and reinforced steel are frequently employed in cutting instruments, chemicals, and coatings that protect them. However, hardness frequently comes at the expense of fragility, which means that these materials can quickly shatter under high stress. In industries where mechanical performance is crucial, such as engineering, building, and production, choosing the appropriate material requires a knowledge of these characteristics. Nanostructures have distinct magnetic characteristics that differ dramatically from bulk substances owing to quantum size effects, surface phenomena, and changed spin dynamics at the nanoscale. These features make magnetic nanoparticles excellent for use in storing information, imaging for medicine, medication delivery, and spintronics technology [18].

Superparamagnetism is among the most prominent magnetic occurrences in nanotechnology. Domains of magnetic attraction align in bulk materials, producing permanent magnetism. However, when the size of magnetic nanoparticles (usually less than 20 nm for materials such as iron oxide) is decreased, they become a single magnetic realm, and fluctuations in temperature can randomize the direction of their magnetism. Consequently, nanostructures can display superparamagnetic behavior, in which they become magnetic only in the midst of a magnetic field from the outside and lose magnetization when the field is removed [19]. This is an extremely desirable property in biomedical applications such as magnetic resonance imaging (MRI) enhancing agents and targeted drug delivery, which require materials to respond to an external magnetic field while retaining minimal residual magnetism in the body. Nanostructures also have increased magnetic distortion, which relates to how magnetic characteristics vary with particle orientation. Nanoparticles' smaller sizes improve surface effects, resulting in increased magnetic anisotropy, which is useful for maintaining magnetization despite temperature variations [20]. This feature is critical in higher-density storage of data, wherein stable and unique magnetic states must exist for accurate information retention. Nanostructured substances also exhibit phenomena such as interchange bias, in which an antiferromagnetic substance interacting with a ferromagnetic nanoparticle produces a unidirectional anisotropy, increasing magnetic stability. Furthermore, quantum-confined phenomena in nanomaterials can alter spin dynamics, resulting in size-dependent variations in magnetic receptivity and coercion [21].

3.3 ELECTRICAL PROPERTIES OF NANOMATERIALS

The confinement of quantum particles has a significant impact on nanomaterials' electric characteristics, notably conductivity. Quantum entanglement happens when the measurements of a nanomaterial are shrunk to a size similar to the electron's wavelength, usually less than 10 nm [22]. Distinctive energy levels, as opposed to the constant energy bands seen in bulk materials, result from the limited electron mobility at this size. This significantly changes the material's electrical characteristics as shown in Figure 3.2. Electrons in bulk materials can travel freely, influencing conductance through established energy bands known as conduction and valence bands. Conversely, when a material shrinks to nanoscale dimensions, the possible states of energy are quantified owing to spatial constraints. This quantization has an influence on the energy dispersion of the electrons, which in turn affects the material's electrical conductivity [23].

The material and its dimensions determine that the quantum confinement affects electrical conductivity. For instance, in semiconducting nanomaterials, the energy gap between the bands of valence and conduction expands as the size of particles reduces. This is because confinement causes electron energy levels to change, necessitating more energy to travel between the valence band to the conduction band. Conductivity may therefore drop with decreasing material size, particularly in materials that include QDs. However, this size-tunable energy gap is useful in applications like as optoelectronics and photovoltaics, where accurate regulation of electrical and

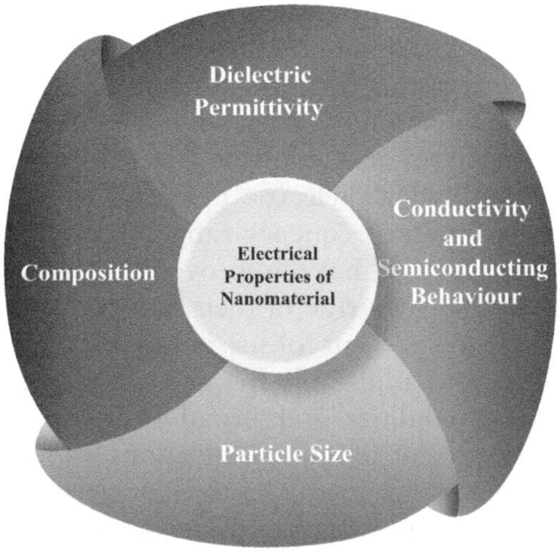

Figure 3.2 Different types of electrical properties of nanomaterials.

optical characteristics is critical. In contrast, quantum confinement in some metallic nanomaterials can increase conductance by forming surface states that allow electrons to flow more easily [24]. In particular, some metallic nanoparticles have higher conductivity than others due to greater electron mobility at the surface, which compensates for the quantum confinement effects in the bulk. Thus, the quantum confinement in nanomaterials influences their electrical characteristics, making them useful in sophisticated electronic and nanotechnological applications. Dielectric permeability refers to a material's ability to retain electrical power when a magnetic field is applied, and it is essential for the operation of capacitors, sensors, and other gadgets. The dielectric permittivity in nanoparticles exhibits different behaviors due to size impacts, surface charges, and the impact of confinement of quantum particles, making them very desirable for applications such as nanotechnology, battery storage, and telecommunications [25].

3.3.1 Dielectric permittivity in nanomaterials

Dielectric permittivity at the nanoscale can differ dramatically from that in bulk materials. This is mostly caused by greater surface area and a higher fraction of atoms on the surface, resulting in stronger polarization effects. Nanostructured materials, including nanoparticles, nanowires, and thin films, exhibit greater dielectric properties than bulk materials, particularly at low frequencies. Nanomaterials elevated surface energy improves the separation of charges, enabling more energy preservation, thus explaining the reason they are commonly employed in high-performance capacitors and supercapacitors [26]. The shape and dimension of nanomaterials impact the frequency responsiveness of dielectric permeability, which defines how the dielectric constant varies with different frequencies of an electric field that is applied. The dielectric permission in bulk materials decreases at greater frequencies due to a delay in dipole reorientation. Nevertheless, nanomaterials, with their huge surface area and quantum effects, can result in a prolonged frequency response, allowing dielectric materials to maintain greater permittivity throughout a wider frequency range. Ceramic nanoparticles such as barium titanate ($BaTiO_3$) have improved dielectric characteristics at low and high frequencies by virtue of their tiny size and optimized surface effects. This makes them excellent for use in miniature electrical devices like embedded capacitors and adjustable frequency filters [27]. Nanomaterials increased dielectric capabilities and broad response to frequency make them very helpful in current telecommunications, wireless communication systems, devices for storing energy, and nanoelectronics, wherein good dielectric efficiency across an extensive frequency range is required.

3.3.2 Conductivity and semiconducting behavior

Materials' conductivity and semiconducting properties vary dramatically at the nanoscale due to their distinct quantum and surface phenomena that develop at these small dimensions. These factors cause nanoparticles to display characteristics that differ significantly from their counterparts in bulk. Conductivity in nanomaterials is strongly dependent on size, shape, and structure. Conductivity in bulk materials is governed by the movement of electrons via the conduction band. However, when a material's properties reduce to the nanoscale (usually less than 100 nm), electron flow is constrained owing to quantum confinement. This results in distinct, quantized energy levels rather than continuous bands, which influence charge carrier movement (electrons or holes). As a result, nanoscale materials' conductivity might be lower or greater [28].

This phenomenon is especially pronounced in semiconductors. The energy distinction between the valence and conduction bands, termed the band gap, expands as the semiconductor's size shrinks. The particle size impact raises the energy needed by electrons to pass from the band of valence to the conduction band, influencing the substance's electrical conductivity. When the size of a material decreases, it becomes harder for electrons to be dynamically stimulated over the bandgap, resulting in less conductivity at ambient temperature. Nanoscale semiconductors, such as QDs, may be created with size-tunable electrical characteristics. The bandgap may be accurately regulated by altering the nanocrystal size, allowing the construction of materials with specified conductivity properties [29]. Nanomaterials' electrical resistivity is substantially impacted by their particle size and arrangement, resulting in unique behaviors in comparison to their bulk equivalents. Recognizing the way these elements influence resistivity is critical for designing materials for particular uses in gadgets, sensors, and power systems.

3.3.3 Particle size

Resistivity frequently rises when nanomaterials' sizes shrink owing to their surface interactions and quantum entanglement. In bulk materials, electrical conduction is largely achieved by the propagation of charged particles (electrons or holes) in an ongoing lattice. Nevertheless, the high surface-to-volume ratio causes more charge carrier scattering at the surface and interfaces when materials are lowered to the nanoscale, usually below 100 nm. This dispersion may result in increased resistance. In metallic nanoparticles, for instance, conductivity is affected by variations in the density of states at the Fermi level as particle size decreases. Nanoparticles below a specific size threshold may switch from metallic to insulation properties, resulting in considerable resistivity variations [30].

3.3.4 Composition

The composition of nanomaterials also has a significant impact on resistance. Various substances have distinct electrical properties, and the addition of dopants or the creation of composite components can significantly change resistance. Doping semiconductors with elements like phosphorus or boron add additional charge carriers, decreasing resistance. In addition, hybrid nanomaterials, which combine multiple kinds of nanostructures (such as metallic and insulator materials), may possess diverse electrical characteristics. The contact properties and interactions between various components in these composites influence overall resistance, enabling the development of materials with tailored conductivity for specific applications [31].

Charge transfer processes are the means by which holes and electrons flow between atoms, molecules, or materials, and can play a crucial role in material magnetic behaviors, particularly at the nanoscale. Understanding these processes is crucial in fields like as electronics, solar technology, and catalysis. At the level of the nanoscale, charge transfer may occur via a variety of different processes [32]. Effective charge exchange is made possible by the movement of electrons along conductivity groups in metals and certain semiconductors. A material, such as a metal, can allow free-moving, delocalized electrons to flow through it when exposed to an electric field. Constant energy bands facilitate electron mobility by allowing them to shift among energy levels with minimal resistance.

Localized states prevent electrons from freely moving throughout numerous disordered or weakly conductive materials, including amorphous semiconductors. As an alternative, people "hop" between different localized sites. This sort of charge transfer is caused by thermal excitation, in which electrons gather sufficient power to migrate across nearby atomic or molecule sites. Hopping, unlike band conduction, is a slower process that is temperature-dependent. At the nanoscale, quantum tunneling provides an important transfer of charge method [33]. During this action, an electron passes past a barrier to energy that it would not be able to cross traditionally. Tunneling is especially essential in nanomaterials, wherein thin insulating boundaries or microscopic gaps among particles allow electrons to travel across without using thermal power. Charge transport in semiconductors, particularly nanoscale systems such as QDs, frequently includes the creation and transfer of electron-hole pairs. When electrons are stimulated from the valence to the conduction band, they leave away holes that operate as positive charge carriers. Conductivity is contributed by both electrons and holes; however, charge mobility can be limited by pair recombination [34].

3.4 COMPARATIVE ANALYSIS OF BULK VERSUS NANOSCALE MATERIALS

Electrical conductivity is a crucial characteristic of materials that greatly affects how they are used for energy storage, electronics, and sensing. Electrical conductivity discrepancies between bulk and nanoscale materials are caused by changes in structural, technological, and thermal properties at different length scales. Electrical conductivity in bulk materials is essentially influenced by the arrangement of atoms in the crystal lattice, as well as the availability of charge carriers like electrons and holes. Temperature, contaminants, and flaws all have an effect on conductivity, which remains typically steady and constant [35]. High conductivity in metals is due to the enormous number of free electrons that may readily flow across the lattice. Conductivity in semiconductors may be modified by doping, which adds more charge carriers.

When materials are reduced to the nanoscale (usually 1–100 nm), their electrical conductivity varies significantly due to a variety of variables. In nanomaterials, a large number of atoms are located at or near the surface. The high surface-to-volume ratio causes enhanced dispersion of carriers of charge at surfaces and interfaces, which may result in higher resistance and poorer conductivity than bulk materials. This impact is more prominent in nanoparticles of metal and thin films. Quantum confinement effects in semiconductors grow more prominent as the size of particles falls. Consequently, the energy levels are quantified, changing the bandgap and causing conductance to vary with size. For example, smaller QDs have higher energy bandgaps and possibly lesser conductivity than bigger ones [36].

Nanoscale materials can experience transitions in phase or changes in structure that affect their electrical characteristics. Certain substances change from metallic to insulating properties as they shrink in size, especially as they reach the nanoscale. Doping and compositional adjustments can fine-tune the conductivity of nanomaterials, resulting in improved device efficiency [37]. Hybrid nanostructures, which combine materials with different electrical characteristics, can also display unique conductance.

The above comparison demonstrates that nanoscale substances have distinct characteristics that differ dramatically from bulk equivalents (Table 3.1). Nanomaterials have a high surface-to-volume ratio, which increases sensitivity and changes their electrical, thermal energy, and mechanical characteristics. Quantum effects become more noticeable,

Table 3.1 Nanomaterials property, comparison between bulk and nanoscale materials

Sl no.	*Property*	*Bulk materials*	*Nanoscale materials*	*Reference*
1	Size	Typically, larger than 100 nm	Ranges from 1 to 100 nm	[38]
2	Surface area	Lower surface-to-volume ratio	Higher surface-to-volume ratio	[39]
3	Electrical conductivity	Generally stable and predictable	Size-dependent, may show increased resistivity due to surface scattering and quantum confinement	[40]
4	Thermal conductivity	Consistent; depends on material composition	Often reduced due to phonon scattering; can vary with size and shape	[41]
5	Mechanical properties	Bulk materials usually exhibit predictable hardness, tensile strength	Increased strength and hardness; often show size-dependent behavior (e.g., Hall–Petch effect)	[42]
6	Optical properties	Defined absorption and emission spectra	Size-tunable optical properties (e.g., QDs exhibit size-dependent fluorescence)	[43]
7	Chemical reactivity	Relatively stable, lower reactivity	Higher reactivity due to increased surface area and unsaturated surface bonds	[44]
8	Magnetic properties	Defined by bulk characteristics	Exhibit unique magnetic behaviors (e.g., superparamagnetism) due to quantum effects	[45]
9	Phase behavior	Stable phase transitions	Size-dependent phase transitions; may exhibit different phases at nanoscale	[46]
10	Doping effects	Consistent response to doping	Greater sensitivity to doping levels, leading to tailored properties	[47]

allowing for tuning of optical and magnetic behaviors. These properties make nanomaterials appealing for a wide range of uses, including gadgets, catalysis, and biological sectors that need precision and customization. Recognizing these distinctions is crucial for scientists and engineers seeking to create novel materials for next-generation technology. Materials' thermal and dielectric characteristics can change greatly depending on their scale, with size, shape, and surface effects predominantly influencing these differences [48]. Knowing these qualities is critical for applications including electronics, energy storage, and thermal control.

Bulk materials have very consistent thermal characteristics on the macroscale, with predictable thermal conductivity and heat capacity. Bulk materials' thermal conductivity is regulated by their crystal structure and phonon transport processes. Because of the free flow of electrons, metals, for instance, usually have high thermal conductivity, whereas insulators have poor thermal conductivity. Thermal characteristics alter dramatically at the nanoscale. The high ratio of surface to volume causes phonon scattering at surfaces and interfaces, which reduces heat conductivity in nanomaterials. This behavior is particularly noticeable in nanoparticles, nanowires, and thin films, wherein phonon transit is limited [49]. When compared to their bulk counterparts, nanoparticles can therefore show reduced thermal conductivity, which is advantageous for thermoelectric and thermal insulation applications. In a similar vein dielectric characteristics vary with scale. Bulk substances often have dielectric constants that are stable due to their molecular structure and composition. Dielectric permittivity in these materials decreases with greater frequency owing to dipole lag after the electric field is applied. Increased surface effects and quantum confinement impact dielectric characteristics at the nanoscale. Due to increased polarization at the surface, nanomaterials frequently have greater dielectric constants than bulk materials [50]. Furthermore, the frequency response of nanomaterials can vary significantly, allowing for a wider range of applications in high-frequency electronics. Development of new materials for energy storage, flexible electronics, and telecommunications is made possible by the distinct electrostatic behavior at the nanoscale.

3.5 APPLICATIONS IN NANOELECTRONICS

Nanomaterials play a revolutionary role in the advancement of battery technology, helping to meet the rising need for high-performance energy storage devices as shown in Figure 3.3. Their unique attributes, including increased surface area, customizable electrical characteristics, and improved electrochemical performance, allow for considerable increases in the density of energy, charging and discharging rates, and total battery efficiency. One of the key uses of nanomaterials in batteries is the creation of improved electrode materials. Silicon nanoparticles are being investigated as anode

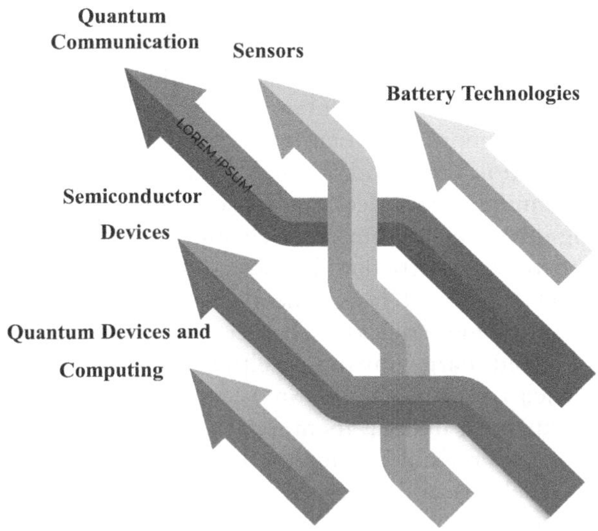

Figure 3.3 Various applications in nanoelectronics.

materials to replace traditional graphite. Silicon has a theoretical capacity of about 4200 mAh/g, which is much larger than that of graphite (about 372 mAh/g). But while cycling, silicon has a tendency to stretch and shrink, which degrades the structure. Through the utilization of silicon nanostructures, such as nanowires or nanoparticles, scientists can adapt to this growth and enhance cycle stability, which will enhance the lifespan and performance of lithium-ion batteries [51].

Graphene and CNTs are two instances of these nanostructures that are used to improve electrode conductivity. Their excellent electrical conductivity allows for quicker electron transit, which is critical for high-power applications. Battery internal resistance may be greatly decreased by adding CNTs or graphene to electrode materials, which will speed up charging and discharging [52]. Nanomaterials are used in solid-state batteries to help create solid electrolytes, which can increase energy density and safety. Nanocrystalline solid electrolytes have higher ionic conductivity than bulk electrolytes, which reduces the possibility of dendrite development, a major problem in traditional liquid electrolyte batteries. Nanomaterials serve an important role in the creation of sophisticated sensors, increasing their sensitivity, selectivity, and response times. Nanomaterials have become crucial parts of many sensor applications, such as environmental monitoring, medical diagnostics, and food safety, because of their special qualities, which include large surface area, programmable electrical features, and catalytic capabilities. The large surface area-to-volume ratio of nanoparticles is one of their main benefits in sensors. This characteristic allows for more contact between the sensor surface and the intended analytes, which improves sensitivity [53]. Metallic nanoparticles, such as gold or silver, can increase signals in surface plasmon resonance (SPR) sensors, allowing them to detect tiny quantities of biomolecules or pollutants in the environment. Additionally, sensor selectivity is improved when nanomaterials are functionalized with particular recognition components, such as aptamers or antibodies. This personalized technique guarantees that sensors respond to specified analytes while reducing interference from other chemicals [54]. CNTs, for example, may be modified with particular ligands and utilized in electrochemical sensors to selectively detect heavy metals or viruses. Nanomaterials also help sensors respond more quickly. Their compact size and unique electrical features enable rapid charge transfer, resulting in faster signal creation. Metal oxide nanoparticles, such as ZnO or SnO_2, have strong reactivity and excellent electron transport, making them ideal for gas sensors. Nanomaterials adaptability makes it possible to employ them in a variety of sensing modalities, such as piezoelectric, optical, and electrochemical sensors. Nanosensors in medical diagnostics may detect illness biomarkers quickly and sensitively, allowing for earlier diagnosis and therapy [55]. Nanomaterials can detect harmful compounds in the air and water, which helps to protect the safety and health of the public.

3.5.1 Fast response times

By improving semiconductor devices' performance, efficiency, and miniaturization potential, nanomaterials have completely changed the industry. As the need for smaller, quicker, and more efficient electronic devices grows, nanomaterials' distinctive characteristics are being utilized to expand semiconductor technology's frontiers. The improved charge transport capabilities of nanoparticles are one of their main benefits in semiconductor devices. Instances of materials that are perfect for interconnects in integrated circuits are graphene and CNTs, which have great carrier mobility and remarkable electrical conductivity. By substituting these materials with conventional metals, resistive losses can be decreased and quicker signal transmission made possible [56].

3.5.2 QDs and light emission

The semiconductor nanocrystals known as QDs, which are usually between 2 and 10 nm in size, have drawn interest due to their special optical characteristics. Because of quantum confinement effects, QDs may produce light at different wavelengths depending on their size. This property is utilized in devices like light-emitting diodes (LEDs) and screens, where QDs offer enhanced efficiency and vivid colors. Highly effective optoelectronic devices may be created because of their capacity to be adjusted for particular wavelengths. The creation of stretchy and flexible semiconductor devices is also made easier by the use of nanomaterials. Material such as organic semiconductors and nanowires may be incorporated into flexible substrates, opening the way for wearable electronics, displays, and sensing. This versatility is critical for the future phase of electronic gadgets, which require lightweight and flexible solutions. Nanomaterials contribute significantly to the functioning of semiconductor devices used in harvesting energy applications [57]. Nanostructured materials, for instance, can increase light absorption and charge separation, thus enhancing solar cell efficiency. In energy conversion systems, nanocomposite films with improved thermoelectric qualities can also be employed to produce power from waste heat.

The development of nanomaterials and nanotechnology is propelling the fast evolution of quantum computers and gadgets. Beyond the constraints of traditional computing, these advancements have the potential to completely transform the way information is processed and stored. Utilizing the ideas of quantum physics, quantum computing enables computations to be completed at previously unheard-of rates. Parallel processing is made possible by the ability of qubits, the basic building blocks of quantum information, to exist in numerous states at once. To create scalable and effective qubits, nanomaterials are essential [58]. High coherence times, for example, are exhibited by superconducting qubits composed of materials such as aluminum or niobium, which are necessary for preserving quantum states. Nanostructured materials are also being used to investigate topological qubits, which are thought to be more durable against environmental perturbations.

3.5.3 Quantum communication

Quantum entanglement is used in quantum communication to transmit data securely. The development of effective quantum repeaters and sources of entangled photons depends heavily on nanomaterials like QDs and photonic crystal structures. These technologies are essential for creating a quantum internet because they provide for safe routes to communicate that are inherently impervious to eavesdropping. By strengthening light–matter interaction and facilitating the production of entangled photon pairs with greater fidelity, the use of nanomaterials improves the functionality of quantum devices for communication. Nanomaterials have also shown great promise in quantum sensing applications, where they may be utilized to produce ultra-sensitive sensors that detect magnetic fields, temperature fluctuations, and gravity waves [59]. Measurements with previously unheard-of accuracy are made possible by quantum sensors, which take advantage of quantum features like superposition and entanglement.

3.6 CASE STUDIES AND RECENT ADVANCEMENTS

Nanomaterials have also shown great promise in quantum sensing uses, where they may be utilized to produce ultra-sensitive sensors that detect magnetic fields, variations in temperature, and gravity waves. Measurements with previously unheard-of accuracy are made possible by quantum sensors, which take advantage of quantum features like superposition and entanglement. For example, diamond-based sensors with nitrogen-vacancy (NV) centers are useful for environmental monitoring and biological imaging as they can identify magnetic fields at the level of the nanoscale [60]. One of the most notable examples is the use of graphene in transistor technology. Graphene, a single layer of carbon atoms arranged in a two-dimensional honeycomb lattice, exhibits exceptional electrical conductivity and high thermal stability. Researchers have developed graphene field-effect transistors (GFETs) that outperform traditional silicon transistors in terms of speed and efficiency. GFETs have demonstrated impressive cut-off frequencies, enabling high-speed operation suitable for RF and microwave applications. Additionally, graphene's flexibility allows for the fabrication of bendable and lightweight electronic devices, expanding the potential applications in wearable technology and flexible electronics [61].

The application of transition metal dichalcogenides (TMDs), including molybdenum disulfide (MoS_2), in transistor production is yet another important breakthrough. TMDs are appropriate for nanoscale applications because they are layered materials with semiconducting qualities that can be stripped down to a monolayer. In comparison to conventional silicon-based transistors, MoS_2 transistors have demonstrated superior on/off ratios, decreased subthreshold swing, and increased electron mobility. TMDs are a strong contender for low-power, high-performance uses in contemporary logic devices because of their features. The goal of recent research has been to combine nanomaterials with current silicon technology to produce hybrid devices that combine the best features of both technologies [62]. For example, adding carbon nanotubes to silicon can improve carrier mobility and lower transistor power consumption. This strategy seeks to push performance boundaries while remaining compatible with existing semiconductor production techniques. Transistor technology has greatly benefited from nanomaterials, resulting in devices that are quicker, more effective, and more versatile. The discovery of graphene and TMD-based transistors demonstrates how nanomaterials can transform the future of electronics. These developments will help devices continue to get smaller as research goes on, making it possible to implement high-performance, low-power electronic applications that are crucial for the digital era [63].

The development of flexible electronic devices has led to new possibilities in a number of fields, including smart fabrics, flexible displays, and wearable technology. Because of their special qualities, which include mechanical adaptability, high electrical conductivity, and compact nature, nanomaterials have played a crucial role in making these advancements possible. CNTs are among the nanomaterials that have made the biggest contributions to flexible electronics. Due to their exceptional electrical conductivity and mechanical robustness, CNTs are perfect for flexible conductive sheets and electrodes. CNT-based translucent conductive sheets, which are lightweight and flexible while maintaining excellent conductivity, have been used by researchers to create adjustable touch screens and displays [64]. In addition to improving user experience, these applications open the door for creative consumer electronics concepts.

Conductive polymers are another family of nanomaterials that have been widely used in flexible electronics. It is possible to process polymers such as polypyrrole and polyaniline into thin films and incorporate them onto flexible substrates. Applications including organic light-emitting diodes (OLEDs), organic photovoltaic cells, and sensors that are flexible require these materials. In contrast to conventional stiff electronics, the use of conductive polymers enables the creation of lightweight, flexible devices that may be manufactured at a cheaper cost.

Additionally, nanomaterials are essential to the development of wearable technologies [65]. Nanomaterial-based flexible sensors that adapt to the surface of the skin to track physiological data including temperature, glucose levels, and heart rate. Flexible electronics have undergone a substantial transformation because of nanomaterials, which have made it possible to create high-performing, lightweight devices that can be incorporated into a variety of applications. Innovations in wearable technology, sensors, and displays have been fueled by the use of conductive polymers, CNTs, and other nanostructured substances. Flexible electronics have the potential to completely transform sectors including electronics for consumers, medical care, and smart textiles as research advances, paving the way for a more networked and versatile future [66].

Nanosensors have developed as strong tools for detecting a wide range of chemicals due to their unique electrical characteristics and great sensitivity. Through the utilization of materials' nanoscale dimensions, scientists have created sensors that can identify even minute quantities of target analytes, opening the door for developments in areas like food safety, medical diagnostics, and environmental monitoring. The size-dependent electrical conductivity of nanosensors is one of the primary characteristics that improve their performance. Because of their one-dimensional and two-dimensional structures, respectively, materials such as graphene and CNTs have exceptional conductivity. These nanomaterials' electrical characteristics alter in response to binding events after they are modified to interact with particular analytes. Nanomaterials of the QD class are also used in electrical sensors [67]. These nanocrystals of semiconductors may be designed for certain detecting tasks owing to their size-tunable optical and electrical characteristics. When stimulated by light, QDs in photonic sensors can produce stronger fluorescence signals, making it possible to identify biomolecules at low concentrations. It is also possible to adjust the electrical response of QD-based sensors such that they react to certain stimuli, including temperature or pH variations.

The application of nanosensors in healthcare has proven transformational, particularly in diagnostics. Nanosensors as well, for instance, may identify biomarkers linked to illnesses like cancer, allowing for rapid diagnosis and individualized treatment regimens. Scientists have created sensors that can swiftly and accurately identify particular proteins in biological samples by employing nanomaterials like gold nanoparticles. This would enable more efficient and prompt medicinal treatments. Detection techniques in a variety of sectors are being revolutionized by nano-sensors, which are boosted by the special electrical characteristics of nanomaterials. They are extremely useful in a variety of applications, from medical diagnostics to environmental monitoring, as their sensitivity to certain analytes [68]. As technology develops, more research into nanosensors should result in more effective, precise, and affordable real-time detection and analysis solutions for important compounds.

3.7 CONCLUDING REMARKS

Significant progress has been made in a number of sectors because of the investigation of nanomaterials and their uses, especially in electronics, sensing, and energy storage. This chapter illustrates how nanoparticles may change physical and electrical characteristics, demonstrating how they have the power to completely transform current technology. Because of their smaller size and larger surface area, nanomaterials have special qualities that set them apart from their bulk counterparts. Improved mechanical strength, thermal stability, and electrical conductivity are some of these qualities, which are essential for use in flexible electronics and semiconductor devices. For instance, materials such as CNTs and graphene have demonstrated remarkable performance in transistors, enabling the development of electrical devices that are quicker and more efficient.

The integration of nanomaterials into sensor technology has enabled the development of highly sensitive and selective devices. Nanosensors can detect trace amounts of chemicals, gases, and biological markers, making them invaluable in healthcare, environmental monitoring, and food safety. The electrical properties of nanomaterials, such as their conductivity changes upon analyte binding, play a crucial role in improving sensor performance. The creation of very sensitive and selective devices has been made possible by the use of nanoparticles in sensor technology. Nanosensors are essential for food safety, environmental monitoring, and healthcare because they can identify minute levels of chemicals, gases, and biological indicators. Enhancing sensor performance is largely dependent on the electrical characteristics of nanomaterials, such as how their conductivity varies when analyte binding occurs. Nanomaterials have improved electrode performance and efficiency in battery technologies, resulting in better energy densities and quicker charge/discharge rates. The development of next-generation batteries that satisfy the requirements of contemporary gadgets and electric cars has shown promise with the integration of tiny silicon particles and conductive polymers.

Future developments in QDs, adaptable electronics, and sophisticated sensing technologies might be greatly aided by ongoing nanomaterials development and research. Nanotechnology is positioned as a foundation for future technological developments due to its special features and capacity to customize materials at the nanoscale. Notwithstanding the notable progress and encouraging uses of nanomaterials, a number of obstacles still need to be overcome to fully use their promise across a range of industries. Research advancement and the smooth integration of nanomaterials into manufacturing processes depend on comprehending and resolving these issues.

The capacity for growth and repeatability of production processes are two of the main issues in the field of nanomaterials. Many synthesis methods work well in the lab but struggle when used in large quantities for commercial settings. For high-quality nanomaterials to be widely used in sectors like electronics, healthcare, and energy storage, dependable and affordable mass manufacturing techniques must be developed. Other difficulties are related to the stability and durability of nanoparticles in practical applications. The performance and dependability of nanomaterials may be impacted by environmental deterioration. Furthermore, extensive research on toxicity, bioaccumulation, and environmental impact assessments is required due to worries about the potential effects of nanomaterial exposure on the environment and human health. Sustainable uses will depend on creating safer nanomaterials and comprehending how they behave in various settings.

Technical and financial difficulties arise when integrating nanomaterials into current systems and technologies. The effective use of nanomaterials in goods is influenced by a number of important aspects, including market acceptability, regulatory frameworks, and compatibility with existing production methods. To overcome these obstacles, producers, researchers, and regulatory agencies will need to work together. Future studies ought to concentrate on a number of important topics, such as scalable synthesis techniques that guarantee the repeatability of nanomaterials with required characteristics developed as a result of innovative synthesis techniques. The creation of hybrid nanomaterials that combine several capabilities for a range of applications is explored because of multifunctional materials.

Utilizing state-of-the-art characterization techniques allows for a greater understanding of the links between nanomaterials' structure and properties. Sustainability promotes research toward environmentally friendly synthesis pathways and methods for recycling to minimize environmental impact. Even though nanomaterials offer fascinating prospects for a range of uses, resolving the related issues is essential to their effective deployment. By emphasizing creative fixes and teamwork, future studies can open the door to the creation of efficient, sustainable, and profitable nanomaterial technology.

The special qualities of nanoparticles permit the downsizing of electrical components while improving their performance, propelling the quick development of nanoelectronics. In the upcoming generation of electronic applications, nanomaterials are expected to be crucial because of the growing desire for devices that are quicker, less, and more environmentally friendly. The creation of outstanding performance transistors is one of the most exciting uses for nanomaterials. TMDs and graphene are two examples of materials with exceptional electrical characteristics, like low power consumption and high electron mobility. Because of these qualities, they may be used to create transistors that function faster than conventional silicon-based devices, opening the door to more sophisticated computer and communications capabilities.

The creation of wearable and flexible electronics, which are becoming increasingly common in consumer goods and medical applications, depends on nanomaterials. Organic semiconductors, carbon nanotubes, and conductive polymers may all be manufactured into thin, flexible films with superior electrical characteristics. Because of their versatility, ordinary gadgets can now have new features because of the development of flexible screens, electronic skin sensors, and smart fabrics. Nanomaterials have great promise for energy harvesting and storage technologies in addition to conventional electrical applications. Nanostructured materials, for example, can improve light absorption and charge separation, increasing the performance of thermoelectric devices and solar cells. Additionally, nanomaterials with better energy densities and quicker charge/discharge rates, such as silicon nanowires and multilayer materials, are being investigated for usage in supercapacitors and contemporary batteries. Researchers are looking at nanomaterials that can build stable quantum bits and improve the performance of quantum systems, ushering in an entirely novel phase of processing. Nanomaterials have enormous potential in new nanoelectronic applications, fueling advances in outstanding performance transistors, versatile devices, harvesting energy, and quantum technologies. The special qualities of nanomaterials will continue to change the electronics industry as research advances, allowing for smarter, more effective, and multipurpose gadgets that satisfy future needs.

REFERENCES

1. Schiavo, L., Cammarano, A., Carotenuto, G., Longo, A., Palomba, M., & Nicolais, L. (2024). An overview of the advanced nanomaterials science. *Inorganica Chimica Acta*, *559*, 121802.
2. Khan, M. Y. A., Elme, N. S., Tahrim, H. M., & Raza, K. (2024). A review on nanotechnology and its impact with challenges on electrical engineering. *Control Systems and Optimization Letters*, *2*(1), 82–89.
3. Fouda-Mbanga, B. G., Onotu, O. P., Olushuyi, C. I., Nthwane, Y. B., Nyoni, B., & Zikhona, T. N. (2024). Application of metallic oxide coated carbon nanoparticles in adsorption of heavy metals and reusability for latent fingerprint detection: a review. *Hybrid Advances*, 100248.
4. Darwish, M. A., Abd-Elaziem, W., Elsheikh, A., & Zayed, A. A. (2024). Advancements in nanomaterials for nanosensors: A comprehensive review. *Nanoscale Advances 6*, 4015–4046.
5. Radsar, T., Khalesi, H., & Ghods, V. (2021). Graphene properties and applications in nanoelectronic. *Optical and Quantum Electronics*, *53*(4), 178.
6. Logeeswaran, V. J., Oh, J., Nayak, A. P., Katzenmeyer, A. M., Gilchrist, K. H., Grego, S., ... & Islam, M. S. (2011). A perspective on nanowire photodetectors: current status, future challenges, and opportunities. *IEEE Journal of Selected Topics in Quantum Electronics*, *17*(4), 1002–1032.
7. Kagan, C. R., Lifshitz, E., Sargent, E. H., &Talapin, D. V. (2016). Building devices from colloidal quantum dots. *Science*, *353*(6302), aac5523.
8. Pokropivny, V., Lohmus, R., Hussainova, I., Pokropivny, A., & Vlassov, S. (2007). *Introduction to nanomaterials and nanotechnology* (pp. 45–100). Tartu: Tartu University Press.
9. Sakthivel, C., Keerthana, L., & Prabha, I. (2019). Current status of platinum based nanoparticles: physicochemical properties and selected applications—a review. *Johnson Matthey Technology Review*, *63*(2), 122–133.
10. Cesaria, M., & Di Bartolo, B. (2017). Nanomaterials: basic concepts and quantum models. In *Nano-optics: principles enabling basic research and applications*, Springer (pp. 43–105).
11. Soni, A. K., & Singh, B. P. (2019). Luminescent materials in lighting, display, solar cell, sensing, and biomedical applications. In *Luminescence-OLED technology and applications*, IntechOpen1.
12. Khan, S., & Hossain, M. K. (2022). Classification and properties of nanoparticles. In *Nanoparticle-based polymer composites* (pp. 15–54). Woodhead Publishing.
13. Meyers, M. A., Mishra, A., & Benson, D. J. (2006). Mechanical properties of nanocrystalline materials. *Progress in Materials Science*, *51*(4), 427–556.
14. Biener, J., Wittstock, A., Baumann, T. F., Weissmüller, J., Bäumer, M., & Hamza, A. V. (2009). Surface chemistry in nanoscale materials. *Materials*, *2*(4), 2404–2428.
15. Jin, R. (2012). The impacts of nanotechnology on catalysis by precious metal nanoparticles. *Nanotechnology Reviews*, *1*(1), 31–56.

16. Rösler, J., Harders, H., & Bäker, M. (2007). *Mechanical behaviour of engineering materials: metals, ceramics, polymers, and composites*. Springer Science & Business Media.
17. Hertzberg, R. W., Vinci, R. P., & Hertzberg, J. L. (2020). *Deformation and fracture mechanics of engineering materials*. John Wiley & Sons.
18. Sudarsan, V. (2017). Materials for hostile chemical environments. In *Materials under extreme conditions* (pp. 129–158). Elsevier.
19. Nocera, T. M., Chen, J., Murray, C. B., & Agarwal, G. (2012). Magnetic anisotropy considerations in magnetic force microscopy studies of single superparamagnetic nanoparticles. *Nanotechnology*, *23*(49), 495704.
20. Cardoso, V. F., Francesko, A., Ribeiro, C., Bañobre-López, M., Martins, P., & Lanceros-Mendez, S. (2018). Advances in magnetic nanoparticles for biomedical applications. *Advanced Healthcare Materials*, *7*(5), 1700845.
21. Sarveena, Shrivastava, N., Singh, M., & Sharma, S. K. (2017). Multifunctional magnetic nanostructures: exchange bias model and applications. In *Complex magnetic nanostructures: synthesis, assembly and applications, Springer* (pp. 225–280).
22. Rossi, M., Militaru, A., Zambon, N. C., Riera-Campeny, A., Romero-Isart, O., Frimmer, M., & Novotny, L. (2024). Quantum delocalization of a levitated nanoparticle. arXiv, Cornell university, https://doi.org/10.48550/arXiv.2408.01264.
23. Goyal, R. K. (2025). Exploring quantum materials & applications: a review. *Journal of Materials Science: Materials in Engineering*, 20.
24. Sherka, G. T., & Berry, H. D. (2024). Insight into impact of size and shape on optoelectronic properties of InX (X = As, Sb, and P) semiconductor nanoparticles: a theoretical study. *Frontiers in Physics*, *12*, 1447997.
25. Yirak, M. G., & Chaujar, R. (2024). Fundamental of emerging nanomaterials. In *Handbook of emerging materials for semiconductor industry, Springer* (pp. 3–24).
26. Kumaravelu, T. A., Jayapalan, R. R., Chang, H. W., Kandasami, A., Vayssieres, L., & Dong, C. L. (2024). Energy storage chemistry: atomic and electronic fundamental understanding insights for high-performance supercapacitors. *Applied Physics Reviews*, *11*(3).
27. Ali, A. I., Alrub, S. A., Hussein, R. K., Son, J. Y., & El Basaty, A. B. (2024). The impact of Gd on structural, morphology, dielectric behavior of $BaTiO_3$. https://doi.org/10.21203/rs.3.rs-4775267/v1.
28. Adul-Rasool, A. A., Athair, D. M., Zaidan, H. K., Rheima, A. M., Al-Sharify, Z. T., & Mohammed, S. H. (2024). 0, 1, 2, 3D nanostructures, types of bulk nanostructured materials, and drug nanocrystals: an overview. *Cancer Treatment and Research Communications*, 100834.
29. Roy, A., Healey, C. P., Larm, N. E., Ishtaweera, P., Roca, M., & Baker, G. A. (2024). The huge role of tiny impurities in nanoscale synthesis. *ACS Nanoscience Au*, *4*(3), 176–193.
30. Pourrahimi, A. M., Olsson, R. T., & Hedenqvist, M. S. (2018). The role of interfaces in polyethylene/metal-oxide nanocomposites for ultrahigh-voltage insulating materials. *Advanced Materials*, *30*(4), 1703624.

31. Lei, W., Liu, G., Zhang, J., & Liu, M. (2017). Black phosphorus nanostructures: recent advances in hybridization, doping and functionalization. *Chemical Society Reviews*, *46*(12), 3492–3509.
32. Adams, D. M., Brus, L., Chidsey, C. E., Creager, S., Creutz, C., Kagan, C. R., ... & Zhu, X. (2003). Charge transfer on the nanoscale: current status. *Journal of Physical Chemistry B*, *107*(28), 6668–6697.
33. Morab, S., Sundaram, M. M., & Pivrikas, A. (2023). Review on charge carrier transport in inorganic and organic semiconductors. *Coatings*, *13*(9), 1657.
34. Royo Valls, M. (2010). Theoretical modelling of electrons and holes in semiconductor nanostructures, [Doctoral dissertation, Universitat Jaume I]. www3.uji.es/~planelle/tesis/Tesis_Miquel.pdf.
35. Fernandez-Garcia, M., Martinez-Arias, A., Hanson, J. C., & Rodriguez, J. A. (2004). Nanostructured oxides in chemistry: characterization and properties. *Chemical Reviews*, *104*(9), 4063–4104.
36. Ahmed, T., & Edvinsson, T. (2020). Optical quantum confinement in ultrasmall ZnO and the effect of size on their photocatalytic activity. *The Journal of Physical Chemistry C*, *124*(11), 6395–6404.
37. Bensebaa, F. (2012). Nanoparticle Technologies: Chapter 7. In *Optoelectronics* (Vol. 19). Elsevier Inc.
38. Zhang, J., Li, Y., An, F. F., Zhang, X., Chen, X., & Lee, C. S. (2015). Preparation and size control of sub-100 nm pure nanodrugs. *Nano Letters*, *15*(1), 313–318.
39. Pozzi, M., Jonak Dutta, S., Kuntze, M., Bading, J., Rüßbült, J. S., Fabig, C., ... & Parak, W. J. (2024). Visualization of the high surface-to-volume ratio of nanomaterials and its consequences. *Journal of Chemical Education*, *101*(8), 3146–3155.
40. Josell, D., Brongersma, S. H., & Tőkei, Z. (2009). Size-dependent resistivity in nanoscale interconnects. *Annual Review of Materials Research*, *39*(1), 231–254.
41. Qian, X., Zhou, J., & Chen, G. (2021). Phonon-engineered extreme thermal conductivity materials. *Nature Materials*, *20*(9), 1188–1202.
42. Naik, S. N., & Walley, S. M. (2020). The Hall–Petch and inverse Hall–Petch relations and the hardness of nanocrystalline metals. *Journal of Materials Science*, *55*(7), 2661–2681.
43. Brkić, S. (2016). Optical properties of quantum dots. *European International Journal of Science and Technology*, *5*(9), 98–107.
44. Xu, L., Liang, H. W., Yang, Y., & Yu, S. H. (2018). Stability and reactivity: positive and negative aspects for nanoparticle processing. *Chemical Reviews*, *118*(7), 3209–3250.
45. Rubel, M. H., & Hossain, M. K. (2022). Crystal structures and properties of nanomagnetic materials. In *Fundamentals of low dimensional magnets* (pp. 183–205). CRC Press.
46. Liu, L., Zhao, R., Xiao, C., Zhang, F., Pevere, F., Shi, K., ... & Sychugov, I. (2019). Size-dependent phase transition in perovskite nanocrystals. *Journal of Physical Chemistry Letters*, *10*(18), 5451–5454.

47. Zandi, O., Agrawal, A., Shearer, A. B., Reimnitz, L. C., Dahlman, C. J., Staller, C. M., & Milliron, D. J. (2018). Impacts of surface depletion on the plasmonic properties of doped semiconductor nanocrystals. *Nature Materials*, *17*(8), 710–717.
48. Guo, Y., Ruan, K., Shi, X., Yang, X., & Gu, J. (2020). Factors affecting thermal conductivities of the polymers and polymer composites: a review. *Composites Science and Technology*, *193*, 108134.
49. Cahill, D. G., Ford, W. K., Goodson, K. E., Mahan, G. D., Majumdar, A., Maris, H. J., ... & Phillpot, S. R. (2003). Nanoscale thermal transport. *Journal of Applied Physics*, *93*(2), 793–818.
50. Kuznetsov, A. I., Miroshnichenko, A. E., Brongersma, M. L., Kivshar, Y. S., & Luk'yanchuk, B. (2016). Optically resonant dielectric nanostructures. *Science*, *354*(6314), aag2472.
51. Duman, M. N. (2024). *Development and synthesis of silicon-based nanomaterials for lithium-ion battery anodes*. Master's thesis, Middle East Technical University.
52. Yang, Z., Tian, J., Yin, Z., Cui, C., Qian, W., & Wei, F. (2019). Carbon nanotube-and graphene-based nanomaterials and applications in high-voltage supercapacitor: a review. *Carbon*, *141*, 467–480.
53. Hassan, R. Y. (2022). Advances in electrochemical nano-biosensors for biomedical and environmental applications: From current work to future perspectives. *Sensors*, *22*(19), 7539.
54. Philip, A., & Kumar, A. R. (2022). The performance enhancement of surface plasmon resonance optical sensors using nanomaterials: A review. *Coordination Chemistry Reviews*, *458*, 214424.
55. Zeng, W., Liu, Y., Mei, J., Tang, C., Luo, K., Li, S., ... & He, Z. (2019). Hierarchical SnO_2–Sn_3O_4 heterostructural gas sensor with high sensitivity and selectivity to NO_2. *Sensors and Actuators B: Chemical*, *301*, 127010.
56. Kolahdouz, M., Xu, B., Nasiri, A. F., Fathollahzadeh, M., Manian, M., Aghababa, H., ... & Radamson, H. H. (2022). Carbon-related materials: graphene and carbon nanotubes in semiconductor applications and design. *Micromachines*, *13*(8), 1257.
57. Liu, Z., Xu, J., Chen, D., & Shen, G. (2015). Flexible electronics based on inorganic nanowires. *Chemical Society Reviews*, *44*(1), 161–192.
58. Priya, S. M., Prasath, S. R., Anitha, S., Jesintha, D. K., & Julie, E. G. (2024). Quantum computing: current advancements and future prospects. In *Emerging trends in IoT and computing technologies* (pp. 132–137). CRC Press.
59. Neuwirth, J., Basset, F. B., Rota, M. B., Roccia, E., Schimpf, C., Jöns, K. D., ... & Trotta, R. (2021). Quantum dot technology for quantum repeaters: from entangled photon generation toward the integration with quantum memories. *Materials for Quantum Technology*, *1*(4), 043001.
60. Schirhagl, R., Chang, K., Loretz, M., & Degen, C. L. (2014). Nitrogen-vacancy centers in diamond: nanoscale sensors for physics and biology. *Annual Review of Physical Chemistry*, *65*(1), 83–105.
61. Norhakim, N., Hawari, H. F., & Burhanudin, Z. A. (2022). Assessing the figures of merit of graphene-based radio frequency electronics: a review of GFET in RF technology. *IEEE Access*, *10*, 17030–17042.

62. Ji, J., & Choi, J. H. (2022). Recent progress in 2D hybrid heterostructures from transition metal dichalcogenides and organic layers: properties and applications in energy and optoelectronics fields. *Nanoscale*, *14*(30), 10648–10689.
63. Akinwande, D., Huyghebaert, C., Wang, C. H., Serna, M. I., Goossens, S., Li, L. J., ... & Koppens, F. H. (2019). Graphene and two-dimensional materials for silicon technology. *Nature*, *573*(7775), 507–518.
64. Li, K., & Zhang, J. (2017). Recent advances in flexible supercapacitors based on carbon nanotubes and graphene. *Cell*, *1*(1), 2.
65. Pode, R. (2020). Organic light emitting diode devices: an energy efficient solid state lighting for applications. *Renewable and Sustainable Energy Reviews*, *133*, 110043.
66. Chiappim, W., Fraga, M. A., Furlan, H., Ardiles, D. C., & Pessoa, R. S. (2022). The status and perspectives of nanostructured materials and fabrication processes for wearable piezoresistive sensors. *Microsystem Technologies*, *28*(7), 1561–1580.
67. Campuzano, S., Yáñez-Sedeño, P., & Pingarrón, J. M. (2019). Carbon dots and graphene quantum dots in electrochemical biosensing. *Nanomaterials*, *9*(4), 634.
68. Thakur, A., & Kumar, A. (2022). Recent advances on rapid detection and remediation of environmental pollutants utilizing nanomaterials-based (bio) sensors. *Science of the Total Environment*, *834*, 155219.

Chapter 4

Carbon nanotube structure, properties, and device applications

Ulfat T. Mir, Sajad Ul S. Dar, Peerzada F. Faroz, A. Yousuf, Khurshed Ahmad Shah, and M. Farooq Mir

4.1 INTRODUCTION

Until the mid-1980s, it was believed that solid carbon existed only in the form of diamond and graphite, both arranged in covalently bonded networks. The diamond is unique among gems as it is made entirely of a single element. It is typically 99.95% carbon. In the remaining 0.05% of the sample, there may be trace elements or atoms that are not essential to the fundamental chemistry of the diamond. Certain trace elements can influence the color and crystal structure of the diamond. Diamonds possess elevated melting and boiling points and are denser compared to graphite. Further, it is transparent and insulating in nature. Diamond formation occurs through the hybridization of carbon's valence electrons to create chemical bonds. Diamond consists of two carbon atoms in each primitive cell, bonded together in a tetrahedral structure. In diamond, each carbon atom establishes covalent bonds with four adjacent carbon atoms, forming tetrahedral units with bond lengths of approximately 1.54 Å. These units arrange themselves in a modified version of the cubic (fcc) crystal structure, with each cube composed of eight carbon atoms [1]. Due to its cubic structure and the symmetrical arrangement of its atoms, diamond crystals can take on different shapes called crystal habits, with the octahedron being the most prevalent form [2]. The cohesive energy per atom in a diamond is about 7.4 eV, and this is the highest of any element. A carbon-based mineral is called graphite. Graphite gets its name from the Greek word graphene, which means "to write," because of its oily texture and black markings. It possesses characteristics of both nonmetals and metals. The electrical and thermal conductivity of graphite is high. Diamond is pliable yet not elastic, extremely resistant to heat, and chemically inert. Graphite is particularly useful in nuclear applications because it minimally absorbs X-rays and neutrons. Its structure consists of layers of carbon atoms organized into six-membered hexagonal rings, stacked in an Aharonov–Bohm (AB) sequence [3]. Graphite has two allotropic forms, "α-graphite" and "β-graphite" which differ in the stacking arrangement of their graphene layers.

DOI: 10.1201/9781003512899-4

In α-graphite, the stacking pattern is ABA (where the 3rd layer aligns directly above the first), whereas in the less energetically stable β-graphite, it follows an ABC sequence (where each layer is offset, creating a three layer repeating pattern).

In 1985, a team of researchers led by Richard Smalley, Robert Curl, and Harry Kroto made the groundbreaking discovery of fullerenes at Rice University and the University of Sussex [4]. They vaporized graphite using intense laser light and helium gas, revealing a mass spectrum showing carbon clusters, notably a strong peak for C_{60} molecules. This led to the proposal of a new carbon allotrope, spherical in shape with 32 faces–12 pentagons and 20 hexagons, resembling a soccer ball. The distinctive soccer ball-like shape seen in various physical objects, such as soccer balls, modern golf balls, architectural designs, and artworks, is attributed to molecules named after architect Buckminster Fuller, known for his geodome designs. The C_{60} molecule, resembling a soccer ball, was dubbed "buckminsterfullerene" or "buckyball" for short. After this discovery, additional carbon-only molecules such as C_{36}, C_{70}, C_{76}, and C_{84} were identified, all classified as a new form of carbon allotrope called fullerenes. These molecules, defined by hexagons and pentagons arranged in a spherical configuration, have garnered attention for potential applications, such as HIV inhibitors and constituents of interstellar space. The distinct geometric characteristics of the novel carbon allotrope extended beyond soccer-shaped molecules.

Sumio Iijima in 1991 found a new way for carbon atoms to arrange themselves—not just in flat sheets like graphite but also in long, hollow cylinders. These tiny tubes, originally called "buckytubes," are now more commonly known as carbon nanotubes or CNTs [5]. He found that when a graphite electrode is hit by an electric arc, the carbon filaments accumulate on the cooler electrode with lengths from one to several microns and diameters in nanometers [6]. These structures bear a resemblance to tubes created by rolling up a sheet of graphite (graphene) or chicken wire. Researchers are actively investigating the unique physical and chemical properties of CNTs in laboratory settings. One notable property is their potential to be just a single atomic layer thick, making them exceptionally thin—approximately of human hair diameter. Due to the bonding nature of carbon atoms, CNTs often resemble rolled-up chicken wire in appearance. An intriguing aspect is that when two nanotubes with slightly different structures are joined together, the junction between them can serve as an electronic device, the behavior of which relies on the specific structures of the tubes.

Efforts are underway to produce CNTs in large quantities with high purity to ensure uniform physical structures [7]. This uniformity would facilitate predicting their behavior, enhancing their utility for potential applications such as nano-sensors. These sensors have the capability to imitate semiconducting materials in microelectronic circuits, perceive minor alterations in electric current, monitor chemical reactions, or measure fluctuations in air pressure and temperature. CNT research is in its early stages, with scientists across chemistry, physics, and material sciences delving into its mysteries and envisioning potential applications.

In recent years, there has been a growing interest in the carbon family, comprising of fullerene [8], carbon nanodots (CQDs) [9], CNTs [10], graphene [11], and graphene quantum dots (GQDs) [12]. This attention stems from the remarkable and unique optical, physical, mechanical, chemical, and electronic characteristics exhibited by these carbon-based materials. For example, semiconducting single-walled carbon nanotubes (semi-SWCNTs) represent a one-dimensional (1-D) quantum-confined forms of carbon allotropes. Their properties, such as near-infrared (NIR) bandgaps, high optical absorption, and ultra-fast charge transport mobility, make them very promising for high-performance photovoltaic devices. Because of exceptional mechanical and chemical stabilities, exceptional electrical and optoelectronic characteristics, and distinct band structure, SWCNTs are also highly attractive options for infrared (IR) detectors in the future [13]. Many IR photodetectors based on single CNTs and CNT films have been successfully shown to date [14]. Recently, a single CNT photodetector was used to create a prototype IR camera [15]. CNT-based photodetectors offer unique advantages for IR detection compared to devices made from other semiconductor materials, including simpler construction, lower fabrication costs, and scalability. In light of this, a wide variety of highly effective photodetectors, such as photodiodes [16], photoconductors [17], and bolometers [18], are rapidly developing.

4.2 STRUCTURE AND PROPERTIES OF CNTS

CNTs are cylindrical formations created by rolling sheets of graphite [19]. The main categories of CNTs are single-walled (SWCNT) and multiwalled (MWCNT), although there are also rarer types such as fullerite, torus, and nano-knot.

SWCNTs are created by rolling a single layer of graphite, into a continuous, cylindrical form, essentially producing extended, wrapped graphene sheets. The diameter of SWCNTs, ranging from 0.4 to 2 nm, is influenced by the synthesis temperature [20].

MWCNTs are made up of multiple SWCNTs arranged in a nested, tube-in-tube configuration. MWCNTs can also be quite long, stretching from one to several micrometers [21].

4.2.1 Electrical properties

The conducting properties can range from metals to semiconductors, having moderate band gaps that depend on their radii and helical structure, along with high strengths and rigidity due to their graphitic and tubular structure.

The electrical properties of SWCNTs are semiconducting or metallic, while MWCNTs can exhibit either metallic or insulating behavior [22]. The electronic properties of SWCNTs can be metallic or semiconducting, depending on their helicity and diameter [23].

4.2.2 Thermal properties

The thermal properties of CNTs have unusual behavior because of their unique size and structure. SWCNT samples that are highly aligned demonstrate thermal conductivity over 200 W/mK at room temperature, with the thermal conductivity of individual nanotubes likely being even greater [24].

4.2.3 Mechanical properties

CNTs are a unique 1-D carbon allotrope that possesses exceptional mechanical properties. The elastic modulus of CNTs has been estimated to range from 1.0 to 5.5 TPa, which is significantly higher than that of steel. A perfect CNT can have a tensile strength exceeding 100 GPa, making it around 100 times stronger than steel cables. The stiffness constant is approximately 800 GPa for SWNTs and around 90 GPa for multiwalled nanotubes (MWNTs) [25].

4.2.4 Optical properties

The optical properties of the CNTs are complex and depend on several factors, such as nanotube geometry, magnetic field, and Zeeman splitting. This study offers a more comprehensive explanation of the optical properties of various CNT types, including details such as the number of absorption peaks and absorption frequencies, which vary based on magnetic fields, nanotube geometry, and Zeeman splitting [26].

4.2.5 Magnetoelectric properties

A magnetic field can alter the physical properties of CNT, such as its energy gap, state degeneracy, and optical excitations. It can also cause a transition from semiconductor to metal in certain types of CNTs. A magnetic field can alter the energy gap, couple different angular momenta, and break state degeneracy, all of which are reflected in the optical excitations. Additionally, this causes oscillatory behavior. When Zeeman splitting is ignored and the magnetic field is aligned with the nanotube axis, periodic AB oscillations appear in the electronic structures. The semiconductor–metal transition happens at a lower magnetic flux due to Zeeman splitting. If a perpendicular magnetic field is present, it can metalize armchair CNTs [27].

Table 4.1 Different properties of carbon nanotubes

Property	*Parameter*	*Range*	*Reference*
Diameter	Armchair (10, 9)	1.32 g/cm^3	[28]
	Zig-Zag (16, 0)	1.33 g/cm^3	[28]
	Chiral (12, 5)	1.41 g/cm^3	[29]
Lattice parameter	Armchair	16.55 nm	[29]
	Zig-Zag	16.53 nm	[29]
	Chiral	16.53 nm	[30]
Structure during equilibrium	Mean diameter	1.3–1.5 nm	[31]
Elastic nature	Young's modulus	1.0–1.27 TPa	[32]
	Tensile strength	Around 100 GPa	[32]
	Armchair	3.37 Å	[33]
Interlayer distance	Zig-Zag	3.40 Å	[30]
	Chiral	3.38 Å	[30]
Thermal property	Thermal conductivity	Around 2000 W/m-K	[34]
	Mean free path	Around 100 nm	[35]
Electrical behavior	Conductance	13.0 (K. ohms)$^{-1}$	[36]
	Current density	1015 A/m^2	[36]

4.3 APPLICATIONS OF CNTS

One of the newest and most advanced technologies is nanotechnology, which offers numerous benefits and the potential to create new materials with notably better qualities. Applications for nanotechnology can be found in various domains, such as energy, environment, sensors, nanomedicine, and energy [37].

4.3.1 Electronic devices using CNTs

CNTs are crucial nanomaterials that can be utilized in electronic equipment to improve their performance [38–48].

4.3.2 Field-emission sources

CNTs can be employed as field-emission sources in electrical devices. This is made possible by applying a potential between the CNT surface and the anode. CNTs can be utilized to fabricate a variety of electronic devices, such as X-ray sources, bright lights, intense light sources, and flat-panel displays [37, 49–54].

4.3.3 Batteries (lithium-ion batteries)

As lithium (Li) has a low electronegativity and can readily donate electrons, it is a useful element with special properties. As a result, it is the ideal option for making batteries that are both lightweight and effective. Despite these benefits, Li's strong reactivity makes it less useful because the metal becomes less efficient. Li ions can be introduced into CNTs to alleviate this issue by combining the applications of CNTs and Li [55–58].

4.3.4 CNTs in supercapacitors

Due to its 1-D nanostructure and exceptional electrical properties, CNTs hold great promise as supercapacitor electrode materials. To attain high voltage and energy density, researchers have recently initiated the development of a high-performance wire-type supercapacitor employing CNTs [59–66].

4.3.5 Drug delivery

CNTs have been utilized as nanocarriers for delivering proteins, genes, and anticancer drugs in chemotherapy through appropriate functionalization. This is due to their distinctive properties, such as high aspect ratios, large surface areas, various surface chemical functionalities, and stable nanoscale dimensions [67, 68].

4.3.6 Gene therapy

Gene therapy and RNA-based approaches have demonstrated considerable promise in the treatment of cancer, and CNTs have the capacity to deliver substantial quantities of therapeutic chemicals, including RNA and DNA, to the targeted disease areas because of their extraordinary flexibility and wire-like structure [69].

4.3.7 Biosensors

Because of their compact size and superior electrochemical properties, CNTs are still very popular as parts of biosensors. In the interim, it is currently well-proven that electrodes based on CNTs have electrochemical attributes that match or surpass the majority of other electrodes. Moreover, it has been shown that chemical manipulation of CNTs is a successful method for adding selectivity to the final biosensors, which have been used in very sensitive DNA recognition [70–72].

4.3.8 Gas and hydrogen storage

As CNTs are empty and cylindrical, they can function as effective metal and gas vessels. The H_2 storage capacity in CNTs has been calculated theoretically using density functional theory. There have been several theories as to how CNTs could soak up molecules of hydrogen, encompassing chemisorption, adsorption at discontinuity (interstitial) locations, and array swelling [73–86].

4.3.9 CNT composite materials

As a result of their remarkable. optical, electrical, and mechanical characteristics, CNTs are utilized to alter and improve the properties of diverse materials including ceramics and polymers [87–89].

4.3.10 CNT-based solar cells

The solar cell industry is paying close attention to CNTs because of their superior features, while Si-based solar cells are becoming increasingly expensive and difficult to further enhance in terms of photovoltaic efficiency [90–92].

4.3.11 Nano tweezers

The development of nano tweezers to facilitate the manipulation and examination of nanostructure elements was made possible by the electromechanical response of CNTs [93, 94].

4.3.12 CNT-based nanosensors

Conventional sensors have the drawback of being anisotropic, whereas CNT-based sensors display isotropic behavior. It is demonstrated that CNT nano-sensors out-sense traditional solid-state sensors by an order of magnitude 30. Compared to conventional sensors, CNT sensors are more compact, have a larger surface area, and have relatively greater electrical and thermal conductivity [95].

4.3.13 X-rays

By aligning a CNT cathode with an anode that has been coated with metal, CNTs can also be utilized to generate X-rays. CNTs can be used to create portable, highly efficient, and low-power-consuming X-ray machines [96].

4.3.14 Photodetector applications

Workhorse devices that translate light into electrical signals are photodetectors. They are vital parts of many devices, such as night vision goggles and the cameras on our phones. A novel class of CNT-based photodetectors has surfaced recently, with promising applications. CNTs are intriguing nanomaterials with special qualities that make them perfect for photodetectors.

CNTs can have a metallic or semiconducting structure. The bandgap, or the energy difference between conducting and nonconducting states, may be accurately regulated in semiconducting CNTs, making it possible to detect particular light wavelengths. From ultraviolet (UV) to IR, CNTs exhibit a great attraction for light throughout a wide range. They can therefore convert light into electrical current more effectively. CNT photodetectors are suited for high-speed applications because of their tiny size and effective charge transfer, which enable them to react to light very quickly.

4.3.14.1 Responsivity

This quantifies the effectiveness with which a photodetector transforms light into an electrical signal. Usually, it is assessed using units such as amperes per watt (A/W) or volts per watt (V/W). A greater responsivity suggests that the photodetector will generate a heightened electrical signal relative to the amount of optical power it receives and is defined as [97].

$$R_V = \frac{Voltage\,generated}{Incident\,optical\,power} \tag{4.1}$$

or

$$R_I = \frac{Current\,generated}{Incident\,optical\,power} \tag{4.2}$$

In our case, we consider these two measures to be equivalent because the internal resistance of the device connects the voltage and the current.

4.3.14.2 Quantum efficiency

This is a gauge of the likelihood that a photon hitting the photodetector will generate an electron–hole pair. It is commonly represented as a percentage and there are two kinds of quantum efficiency [97].

4.3.14.3 External quantum efficiency

This is the proportion of electrons gathered per second compared to the number of photons hitting the detector per second.

$$EQE = \frac{Photocurrent/Charge\ of\ one\ electron}{Incident\ optical\ power/Energy\ of\ one\ electron} \tag{4.3}$$

4.3.14.4 Internal quantum efficiency

This is the relationship between the rate of electron–hole pair production per second and the rate of photons striking the detector per second.

$$IQE = \frac{EQE}{1 - reflection - transmission} \tag{4.4}$$

In addition to responsivity, external quantum efficiency (EQE), and internal quantum efficiency (IQE), other key performance metrics for photodetectors operating at zero bias include short-circuit current (I_{SC}), open-circuit voltage (V_{OC}), fill factor (FF), and power conversion efficiency (η).

The FF and power conversion efficiency are determined by [97]:

$$FF = \frac{I_M V_M}{I_{SC} V_{OC}} \tag{4.5}$$

$$\eta = \frac{I_M V_M}{P_{in}} \tag{4.6}$$

where I_M and V_M represent the levels of photocurrent and photovoltage at which the output power achieves its peak, and P_{in} denotes the incident power density.

4.3.14.5 Noise equivalent power

Noise equivalent power (NEP) is the minimum optical power that a photodetector can detect with a signal-to-noise ratio of 1. It quantifies the sensitivity of the photodetector. A decreased NEP suggests that the photodetector is capable of sensing fainter signals [97].

$$NEP = \frac{Noise\,spectrum}{Responsivity} \tag{4.7}$$

where the noise spectrum has units of $V / \sqrt{Hz}$ for a responsivity RV (responsivity value) so that the NEP has units of $W / \sqrt{Hz}$. This figure represents the minimum power detectable above the background noise.

4.3.14.6 Detectivity

This metric combines both the responsivity and noise characteristics of a photodetector, serving as a measure of its overall performance. It measures the signal-to-noise ratio (SNR) that a photodetector can achieve at a given optical power level. Detectivity is typically measured in units of $cm\sqrt{Hz}\,/\,W$ (Jones). An increased detectivity suggests that the photodetector will possess an improved SNR, enabling it to detect more delicate signals. Another metric that standardizes the NEP by the active area of detector A is known as specific detectivity [97]:

$$D = \frac{\sqrt{A}}{NEP} \tag{4.8}$$

4.3.14.7 Response time

This denotes the duration for the output signal of a photodetector to increase from 10% to 90% of its ultimate value following a sudden alteration in light intensity. This parameter holds significance in applications like optical communication and high-speed imaging. The speed of response can be measured using either the rise or fall time of the photosignal (τ) or the cut-off frequency (f_c). These two measurements are related to each other as [97]:

$$f_c^{-1} = 2\pi\tau \tag{4.9}$$

4.3.14.8 Active area

This refers to the region of the photodetector that responds to light. A greater active area can capture more light, enhancing the sensitivity of the photodetector [97].

The carbon allotropes extensively investigated for photonic applications are 2-D graphene and 1-D CNTs [98, 99]. These allotropes possess advantageous properties including robust environmental resistance, wide-ranging optical absorption, and exceptional carrier mobility [18, 100–102]. Comparatively speaking, individual CNTs may move up to 79,000 cm^2 at room temperature; CNT networks typically exhibit decreased mobility, usually ranging from 10 to 100 $cm^2/(V{\cdot}s)$, as a result of factors such as inner-tube Schottky barriers and various morphological factors, leading to slower operational speeds [103–107]. However, graphene's ability to respond to light is significantly constrained due to its poor optical absorption and rapid recombination of photo-generated carriers [108, 109]. Scientists have created new ultra-thin structures by combining SWNTs and graphene. These new materials, linked together either by strong chemical bonds [110–112] or weaker attractive forces [113–115], outperform the individual components. Researchers are now working on ways to precisely control the microscopic structure of these graphene-SWNT hybrids. The recent successes of "rebar graphene" [111] and "rivet graphene" [112] show that it is possible to tailor the properties of these hybrids for specific applications.

Photodetectors based on graphene have been studied extensively. Even though intrinsic graphene-based high-speed devices have been used at several tens of GHz, the considerably low responsivity (~10^{-3} A/W) [115] continues to be a barrier to practical use. The most effective method for increasing photoresponsivity has been photogating graphene; photoresponsivity as high as $10^7\ A/W$ has been shown using a quantum dots layer. However, it always results in a weakened reaction time [116]. An excellent photodetector with well-rounded performance was recently accomplished by creating graphene-SWNT van der Waals heterostructures on a SiO_2/Si substrate. The hybrid device combining graphene and SWNT demonstrated a photoconductive gain of 10^5 and achieved a rapid response time of around 100 µs. This suggests its potential suitability for light harvesting applications requiring expansive sensing coverage, elevated photoresponsivity, and swift operational speeds [114]. Such hybrid nanostructures' interfacial charge transfer kinetics are also determined by combining photocurrent and Raman analysis methods for spectroscopy [117].

Due to their numerous applications, flexible electronics with optoelectronic characteristics incorporated on bendable plastic substrates have garnered a lot of interest in wearable technology, smart windows, and paper displays [17, 118–122]. Unlike rigid devices, the materials needed to construct flexible devices must be inherently stiff, bendable, and strong enough to endure high strain levels [123]. Surprisingly, graphene-SWNT complexes display remarkable electrical and optical conductivity, mechanical flexibility, and stretchability, making them attractive options for flexible optoelectronic systems [115–125].

4.4 TYPES OF CNT-BASED PHOTODETECTORS

Photodetectors based on CNTs represent a promising and innovative category with distinct advantages over conventional counterparts. CNTs possess exceptional electrical and optical characteristics that make them particularly suitable for photodetection applications. One notable feature is their ability to exhibit either metallic or semiconducting behavior based on their chirality, with their band gap being adjustable by altering their diameter. Consequently, CNT-based photodetectors can effectively detect a broad spectrum of light wavelengths.

Various types of photodetectors utilizing CNTs exist, each carrying its own set of strengths and weaknesses. Below are several prevalent variants:

4.4.1 Photoconductor

Photoconductive detectors, also referred to as photoresistors or light-dependent resistors (LDRs), are constructed from materials whose conductive properties alter upon exposure to light. When illuminated, these detectors generate electron–hole pairs, leading to an enhancement in material conductivity. Electron–hole pairs generated optically are segregated within the channel because of electric fields produced by an applied source–drain bias. A significant challenge with SWCNT photoconductors is that the binding energies of photo-induced electron–hole pairs are substantial, typically reaching several hundred meV [126], surpassing the thermal energy at room temperature. As a result, the efficient separation of electron–hole pairs requires a robust electric field, which ultimately limits the performance of the detector [18]. While photoconductive detectors offer simplicity and cost-effectiveness in fabrication, they generally exhibit lower sensitivity compared to photodiodes or phototransistors. It is important to highlight that impurities present on the s-SWCNT significantly impact the dissociation and binding of excitons [127, 128].

One potential method to improve the separation of electron–hole pairs involves incorporating CNTs into specific polymers. Pradhan et al. [120, 129] conducted pioneering research in this area. The researchers made IR sensors using a mix of SWCNTs and polycarbonate. These sensors were more sensitive to light than heat. When compared to sensors made only of CNTs, the new sensors were 10 times more effective in detecting IR light.

Subsequent research revealed that other polymers, like (6,6)-phenyl-C71-butyric acid methyl ester (PC71BM) [130, 131] and poly[3-hexylthiophene] (P3HT) [120, 132], can also form heterojunctions with CNTs. These materials can be combined with tiny, electrically conductive tubes to make a special type of material called a bulk heterojunction [130, 131]. This material has many tiny junctions between the two different materials throughout its structure. The electrons and holes that are created at the junctions in these materials help to produce an electrical current when an electric field is applied. For example, Wu et al. [17] made a photoconductor using a mix of a semiconductor, SWCNTs, and a polymer called P3HT. This photoconductor could detect very small amounts of light with a detectivity (D^*) of $2.8\times10^8\ cmHz^{\frac{1}{2}}/W$ [120].

The researchers used a special technique called capacitive photocurrent spectroscopy to study the photocurrent in individual CNTs [133] and films made of CNTs [134]. This technique can tell the difference between the photoconductive effect and the thermal effect. The study found that in films of CNTs, electrons and holes can be separated when a strong electric field is applied. However, the photoconductivity caused by the separation of electron–hole pairs is typically much less significant than the photoconductivity resulting from oxygen desorption [121] and bolometric effects [18].

In 2001, Fujiwara et al. [122] documented the occurrence of photoconductivity in a semiconducting single-walled carbon nanotube (s-SWCNT) measuring 1.4 nm in diameter [135]. They identified two distinct peaks in the photoexcitation spectrum, located at approximately 0.7 and 1.2 eV, which corresponded to the first and second sub-bands of the s-SWCNT. In

2012, Lu et al. [17] created a photoconductive sensor by blending poly(3-hexylthiophene) polymer with a film of s-SWCNTs, aiming to separate excitons at the interface of the heterojunction [136]. Under a bias voltage of 10 volts, the detector achieved its highest responsivity and detectivity (D), measuring 2.2 mA/W and $2.3\times10^8\ cmHz^{\frac{1}{2}}/W$ (Jones), respectively. The new detector was able to turn on and off very quickly, taking only about 0.6 and 1.4 ms. This is much faster than thermal detectors. Photoconductive detectors are also better at detecting very small amounts of light. However, the new detector needs to be operated at a high voltage to work properly. This is because the excitons in the SWCNTs have a higher binding energy than the thermal energy at room temperature. Operating at a high voltage increases the power consumption and the dark current of the device.

In 2014, Park et al [123] fabricated phototransistors utilizing a combination of s-SWCNTs and C_{60} [137]. The new device is different from photoconductive devices in that the electrons get trapped in a material called C_{60} after the excitons are separated into electrons and holes by an electric field. By changing the voltage and the size of the device, the researchers were able to make the detector much more sensitive. It could detect light that was 2000 to 40,000 times weaker than before. The maximum detectivity of the device was more than $10^9\ cmHz^{\frac{1}{2}}/W$ [138].

Wang et al. [165] reported a photoconductor utilizing a solitary layer of horizontally aligned SWCNTs [125]. The device had an array of these nanotubes arranged across a silicon trench that was 40 μm wide. Electrodes were connected to both ends of the CNTs. When the device was exposed to IR light with a power of less than 4 mW, its resistance decreased by 22.86%. The researchers measured the response time to be about 0.5 ms. They believe that the photoresponse is caused by the separation of electrons and holes along the CNT channels when an electric field is applied [125]. Responsivity or detectivity measurements were not provided for these devices. The main problem with the current device is that there are not many CNTs in the film. In fact, there is only about one tube per micron. This means that the device does not absorb much light. To improve the performance of the device, the researchers need to find ways to make thicker films of CNTs that can absorb 100% of the light. These films would need to be at least 60–100 nm thick.

4.4.2 Bolometer

Bolometers consist of resistive components made from active materials characterized by low heat capacity and a high-temperature resistance coefficient [18, 135]. When bolometers absorb thermal radiation, their resistance undergoes significant alterations. The ability of the detectors to respond quickly depends on the sensitivity of the active material to temperature changes and the overall temperature of the environment. Typically, the detector's active material is suspended to isolate it from the surrounding thermal effects. In 2006, Itkis et al. presented a bolometer utilizing a suspended film of s-SWCNTs with varying thicknesses (40, 100, and 1000 nm) within a vacuum setting [127]. The device's ability to detect light decreased as the film became thicker when exposed to light with a wavelength of 940 nm and a power of 0.12 W. At very cold temperatures of 4.2 K, the device could produce a voltage of up to 1000 V/W, and it took about 50 ms for the device to respond to the light. The temperature coefficient of resistance (TCR) of SWCNTs film can be boosted additionally by decreasing the thickness, enhancing the purity of s-SWCNTs, and chemically altering the s-SWCNTs. Apart from s-SWCNTs as active components [128, 138, 139], multiwalled CNTs [140, 141], and hybrids of s-SWCNTs with polymers [139, 142] have attracted attention in research. The detector can detect light across a wide range of IR wavelengths, unlike detectors that are limited by the material's bandgap. Bolometers are used for measuring temperature, monitoring industrial processes, and serving many other applications [138]. To measure the sensitivity of a bolometer, its responsivity is calculated using a specific formula [143]:

$$R_V = \frac{\Delta V}{\Delta P} = \frac{\iota R \alpha \eta}{\sqrt{G^2 + \omega^2 C^2}} = \frac{\iota R \alpha \eta}{G\sqrt{1 + \omega^2 \tau^2}} \tag{4.10}$$

where ΔP is the incident power on the active detection area, ΔV is the corresponding voltage change, η is the absorption efficiency, R is the resistance of the active material, $\alpha = (dR/dT)/R$ is the TCR, G is the thermal conductance to the substrate, C is the heat capacitance of the detector, and ω is the modulation frequency. The response time of the bolometer can be calculated using

$$\tau = \frac{C}{G} \tag{4.11}$$

Efficient bolometers, as indicated by Equations 4.10 and 4.11, rely on substantial optical absorption, a high-TCR, minimal heat capacity, and effective thermal isolation from the surroundings (resulting in a low thermal conductance, G).

In the last 10 years, there has been much progress in the development of bolometric detectors made from CNTs. These detectors are used to detect IR radiation. There are three main types of CNT-based IR detectors: (1) SWCNT films [136, 138, 143, 144], (2) composites of SWCNTs with polymers [145–150], and (3) MWCNT bundles or films [97, 151–156]. These devices usually take about a few hundredths of a second to respond, and their typical detectivity ranges from approximately $10^6 - 10^7\ cmHz^{\frac{1}{2}} / W$ (Jones). They operate across wavelengths from the NIR (around 0.94 μm) to the long-wavelength IR region (up to about 20 μm). However, the ability of current

CNT-based bolometers to detect IR radiation is still much lower than that of commercial bolometers by 1–2 orders of magnitude, which is about $10^8\ cm\,Hz^{\frac{1}{2}}\ /\ W$ (Jones) [157]. When using SWCNT-polymer composites instead of SWCNT films, the TCR has increased significantly, from about 0.1% per kelvin to around 10% per kelvin [151].

4.4.3 Photothermopiles

The photothermoelectric (PTE) detector, also referred to as the photothermopile, stands as a commonly utilized form of thermal detection technology [158]. In contrast to the bolometer, a PTE detector has the potential to operate without any current or voltage, thus requiring no external power consumption. Although they exhibit lower sensitivity and speed when compared to photonic detectors, CNT-based PTE detectors offer an exceptionally wide-ranging response, spanning across the electromagnetic spectrum from UV and visible light to IR and the terahertz (THz) region [158]. Their advantages lie in their ability to detect long-wavelength radiation, making them highly promising for THz detection at room temperature.

The PTE detector operates based on the Seebeck effect. When light strikes the device, it generates heat, resulting in a temperature gradient along the channel. This gradient, in turn, produces a photovoltage (or photocurrent) across the device [159, 160].

The thermovoltage is defined as

$$\Delta V = -\int S \nabla T dx \tag{4.12}$$

where S is defined as the Seebeck coefficient of active material, and ∇T is the temperature gradient of the detector channel. While S is an intrinsic property of the material, achieving significant temperature gradients demands that the device be properly thermally isolated from its surroundings, akin to what is required for a bolometric detector [151]. According to Equation 4.12, the device must generate a temperature gradient, which results in three primary device designs: In the initial design, the device exhibits symmetry, with only one of the two contacts being exposed to illumination. This setup induces a temperature gradient between the two electrodes. In the second design, a *p–n* junction is formed within the channel, and upon illumination, this *p–n* junction generates a temperature gradient between itself and the contacts. In the third configuration, the film remains consistent throughout, yet distinct metals are employed for the electrodes, while the device receives illumination from all directions [151]. In a symmetric device where illumination occurs at one contact [161], the thermovoltage can be expressed as

$$\Delta V = \Delta T \left(S_{CNT} - S_{metal} \right) \tag{4.13}$$

Here, ΔT refers to the temperature difference between the two electrodes, and S_{metal} indicates the Seebeck coefficient of the metal electrode. Generally, this factor is negligible compared to that of the active material (in this case, the CNT material) and can often be overlooked in a high-performing device. In the case of a *p–n* junction, the situation differs as

$$\Delta V = \Delta T\left(S_p - S_n\right) \tag{4.14}$$

Here, S_p and S_n represent the Seebeck coefficients for the *p*-type and *n*-type CNT materials, respectively. Given that the Seebeck coefficients of *p*-type and *n*-type materials have opposite polarities, it is clear from these expressions that a *p–n* junction configuration provides advantages, provided high Seebeck coefficients can be achieved for both *p*-type and *n*-type doping, along with the generation of comparable temperature gradients [151]. (Just like with the bolometric detector, the device's response time is determined by $\tau = \frac{C}{G}$.)

In a simplified situation where Johnson noise is the main contributor to detector noise, the detectivity of a CNT-PTE detector depends upon the thermoelectric figure of merit ZT [162], according to $D^* \sim \sqrt{ZT}$, where

$$ZT = \frac{S^2\sigma}{\kappa} \tag{4.15}$$

where κ represents the thermal conductivity of the CNT film. Therefore, enhancing the effectiveness of PTE detectors necessitates enhancing the comprehensive thermoelectric characteristics of the active material [151].

CNT-PTE detectors have been fabricated utilizing extensive CNT films, such as CNT networks [163–165], aligned CNT films [13, 14, 83, 166], and CNT-polymer blends [166]. The device configurations vary, encompassing *p–n* junctions [14, 83, 163, 164], disparate metal electrodes [13], and dissimilar thermal connections [83, 167]. It is evident that the response time of these detectors typically tends to be sluggish, typically ranging from milliseconds to seconds, owing to their dependence on thermal dissipation. Presently, the detectivity of CNT-PTE detectors remains relatively low, standing at less than $10^7\ cmHz^{\frac{1}{2}} / W$ (Jones). While this figure is on par with many CNT bolometers, it falls significantly short when compared to the top-performing photonic detectors operating in the NIR range. As previously noted, the strength of these detectors lies in their wide-ranging response, enabling their use in scenarios where photonic detectors typically exhibit limited responsivity. An illustrative case is the THz range, known for posing considerable challenges to photonic devices. Two archetypes of THz PTE detectors employing CNT films have already been created [13].

Possible approaches to improving the speed and sensitivity of CNT-PTE detectors include (1) enhancing the thermoelectric characteristics through the investigation of factors such as CNT alignment, diameter distribution, quality of junctions, density, and more; (2) developing techniques to produce suspended ultrathin films to reduce heat capacity and improve response time; and (3) formulating strategies for voltage amplification utilizing multiple junctions [151].

4.4.4 Photodiodes

A recently developed photodiode, known as a CNT-based photodiode or CNT photodiode, employs CNTs to absorb light and generate electrical current [168]. The evolution of CNT photodiodes has roughly followed the three phases observed in the development of photoconductors: (1) photodiodes consisting of individual semiconducting SWCNTs, (2) photodiodes utilizing arrays or films of semiconducting SWCNTs, and (3) photodiodes employing heterojunctions of semiconducting SWCNT films and polymers. Since 2003, the development of CNT photodiodes has progressed from individual SWCNT-based devices to hybrid structures integrating semiconducting SWCNT films with polymers. This advancement has involved diverse device designs including field-effect transistors (FETs) [168], *p–i–n* diodes [17], barrier-free bipolar diodes (BFBD) [169], and heterojunctions [130].

The advancement of CNT photodiodes addresses two primary challenges: (1) developing a device that effectively segregates electron–hole pairs into liberated carriers and (2) enhancing the device's absorption cross-section. In recent times, notable gains have been made, leading to a substantial increase in responsivity from approximately picoamperes per watt to approximately amperes per watt. To date, the most effective CNT-based photodetector utilizes an SWCNT/C_{60} planar heterojunction (PHJ) configuration, exhibiting an EQE of 12% and achieving a detectivity close to around $10^{12}\ cmHz^{\frac{1}{2}}/W$ (Jones) [131]. Like CNT photoconductors, the absolute absorption of an individual CNT photodiode is inadequate for the majority of photodetector applications [131]. Applications of this nature necessitate the utilization of expansive CNT films to create photodiodes that better mimic real-world conditions. To create a photodiode using SWCNT film that possesses both high speed and sensitivity, two issues must be addressed. The initial challenge involves addressing the purification of SWCNTs based on their electronic type. The second aspect involves enhancing the separation competence of electron–hole pairs within bulk SWCNT films.

The BFBD architecture proves to be an optimal design for creating individual CNT diodes. Peng et al. [167] successfully applied this method first to SWCNT networks [132] and subsequently to aligned SWCNT arrays [170]. The IR photovoltaic detector utilizing the semiconducting SWCNT array demonstrated significant success, outperforming individual CNT devices by a considerable margin. When exposed to IR illumination at a wavelength of 785 nm, the short-circuit photocurrent showed a linear increase as the incident power density increased. Similarly, the open-circuit photovoltage displayed a logarithmic increase with growing light intensity, demonstrating typical behaviors expected from a photodiode. The peak responsivity and the D^* values (calculated based on the actual area of CNTs) were approximately $6.58\times10^{-2}\ A/W$ and approximately $1.09\times10^{7}\ cmHz^{\frac{1}{2}}/W$ (Jones), respectively. This represented the first experimental measurement of the detectivity of a CNT photonic detector, exceeding that of most CNT bolometers. Performance improvements could be realized by increasing the density of semiconducting SWCNTs in the channel; it was estimated that a CNT density of 60 tubes per micron could result in nearly a tenfold enhancement in the device's D^*, reaching around $10^{8}\,cmHz^{\frac{1}{2}}/W$ (Jones).

Multiple research teams have successfully developed photodetectors by forming heterojunctions between semiconducting SWCNTs and polymers, employing various charge-accepting materials in the process [123, 171–175]. The performance of these detectors has seen a significant enhancement compared to both the SWCNT array detector and CNT-polymer photoconductors. Of these, detectors based on semiconducting SWCNT/C_{60} planar heterojunctions (PHJs) have notably stood out for their success and extensive research exploration [176, 177].

In 2009, Forrest et al. [170] introduced a photovoltaic detector operating from visible to NIR wavelengths, employing a semiconductor-SWCNTs/C_{60} planar heterojunction (PHJ) [174]. The detector exhibited outstanding performance, featuring a D^* greater than $10^{10}\ cmHz^{\frac{1}{2}}/W$ (Jones) across the wavelength spectrum from 400 to 1400 nm, along with a response time of approximately 7.2 ns. At a wavelength of 1155 nm, the EQE was measured at 2.3%, while the IQE stood at 44%. In 2012, Bindl et al. [195] created an even more effective IR detector using semiconducting SWCNTs and C_{60}, achieving a detectivity (D*) of up to $6\times10^{11}\ cmHz^{\frac{1}{2}}/W$ (Jones) [130]. The maximum values for EQE and IQE were increased to 12.9% and 91%, respectively. In this device, the active layer comprised semiconductor-enriched SWCNTs encased in poly(9,9-dioctylfluorene) (PFO). Subsequent research conducted by this team demonstrated that enhancing the EQE of the detector could be achieved through the removal of excess PFO and the appropriate selection of CNT chirality. The peak EQE of approximately 34% was attained in a photovoltaic detector utilizing the monochiral (7,5)-SWCNT/C_{60} planar heterojunction, with the SWCNT layer thickness being approximately 7 nm [178].

Scientist developed a new type of CNT diode in 2013. This diode had a very low leakage current and was able to effectively rectify electrical signals. The researchers used a split-gate and asymmetric contact geometry to achieve these results [179]. The diode had a leakage current of only 2×10^{-14} amps and an ideality factor of 2.01.

Peng et al. [167] developed a new type of IR photodetector in 2016. This photodetector used high-purity SWCNT networks and was able to detect a wide range of IR wavelengths. The devices had a peak voltage responsivity of $1.5\times10^{8}\ V/W$ and a high detectivity of up to $1.25\times10^{11}\ cmHz^{\frac{1}{2}}/W$ (Jones) [180].

4.4.5 Phototransistors

In a typical transistor, the amount of current flowing between the source and drain is controlled by a gate voltage. This allows the transistor to be turned ON or OFF. If the gate voltage is changed by light, the current flowing between the source and drain will also change. This change can be measured electrically [151]. This type of photodetector is called a phototransistor. In these devices, the main focus is on how the transistor's behavior changes when it is exposed to light. This change is often seen as a shift in the transfer curve of the transistor [151].

The development of CNT phototransistors is currently less advanced compared to other devices pertinent to practical applications. CNT transistors were initially employed to investigate the fundamental photophysics of individual CNTs; however, there have been limited studies showcasing actual phototransistor functionality. The light exposure can change the threshold voltage of the transistor, causing it to shift. Sometimes, this happens because of charge traps in the material underneath the transistor. This phenomenon is called photogating. However, managing and incorporating such occurrences into practical applications proves challenging. More hopeful are methods that deliberately incorporate light-sensitive elements near the CNTs, which act as efficient gates. This method has shown an exceptionally high responsivity of $10^4\ A/W$, along with spectral selectivity achieved through well-designed chromophores. However, it has solely been employed on individual CNT devices within the visible spectrum thus far. A similar method was recently showcased involving the deposition of C_{60} layers on CNT thin film transistors [181], achieving a detectivity of $10^9\ cmHz^{\frac{1}{2}}/W$ (Jones) in the near-IR spectrum. Additionally, quantum dots [182] with visible light absorption have been employed in a similar manner.

In 2012, researchers developed a new type of phototransistor that used CNT/titanium dioxide core-shell nanowires. This phototransistor exhibited a high photogain of 1.4×10^4, with a response time of 4.3 ms and a recovery time of 10.2 ms in the UV–visible detection range [183]. In 2013, Bao et al. [196] fabricated a new type of photogating transistor that used SWCNTs/C_{60} hybrids. This phototransistor had an improved responsivity of approximately 80A/W and a detectivity of $1.17\times10^{9}\ cmHz^{\frac{1}{2}}/W$ (Jones) at a wavelength of 1200 nm [184].

In 2017, Adinolfi and Sargent [193] developed a new type of photovoltage field-effect transistor that used silicon and lead sulfide quantum dots. This transistor had a high gain, high speed, and a large gain-bandwidth product at a wavelength of 1500 nm [185]. The device was still limited by the problem of charge trapping. This problem affected both the conversion of light into electricity and the movement of charges in the active part of the transistor.

In 2019, scientists developed a new type of phototransistor in 2019. This phototransistor was made of a CNT network and all-inorganic perovskite $CsPbBr_3$-quantum dots. The phototransistor showed a high responsivity of $5.1\times10^7\ A/W$ and a specific detectivity of $2\times10^{16}\ cmHz^{\frac{1}{2}}\ /W$ (Jones) at a wavelength of 516 nm [186].

In 2020, researchers achieved a giga-gain exceeding in a CNT/P3HT hybrid phototransistor under 530 nm illumination. The high gain resulted from the trapping of negative charges near the CNTs, creating an integrating detector. This detector could sense as little as 490 attowatts of light and distinguish as few as 8–13 photons per nanotube at room temperature [187].

4.5 FUTURE PERSPECTIVES

Research on the photodetector effects of CNTs has a promising future due to their unique properties and the growing demand for advanced photodetector technologies.

CNTs offer exceptional sensitivity due to their high surface area, tunable band gaps, and excellent electron mobility. Future research may focus on optimizing these properties to develop photodetectors with ultra-high sensitivity for applications in low-light environments, such as night vision or space exploration.

CNTs can be engineered to absorb light across a wide spectrum, from UV to IR. Research is likely to continue exploring ways to tailor CNTs for specific wavelength ranges, potentially leading to the development of broadband photodetectors useful in spectroscopy, medical imaging, and environmental monitoring.

The flexibility and mechanical strength of CNTs make them ideal for integration into flexible and wearable electronics. Future research could lead to the development of photodetectors that are lightweight, bendable, and even stretchable, opening up new possibilities for wearable sensors, smart textiles, and flexible displays.

Combining CNTs with other nanomaterials, such as graphene, quantum dots, or perovskites, could lead to hybrid photodetectors with enhanced performance. Research in this area might focus on optimizing these combinations to achieve higher efficiency, faster response times, and lower power consumption.

As research progresses, there will likely be a focus on developing cost-effective and scalable production methods for CNT-based photodetectors. This could involve advances in chemical vapor deposition (CVD) techniques,

Table 4.2 Comparison of various parameters of CNT-based photodetectors and photodetectors based on other material

Feature/parameter	*CNT-based photodetector*	*Silicon photodetector*	*Germanium photodetectors*	*III–V compound photodetectors*
Material	CNTs	Silicon	Germanium	For example, InGaAs
Responsivity	High (up to $10^6 A/W$)	Moderate (up to ~$0.6 A/W$)	High (~$1 A/W$)	High (~$1 A/W$)
Speed	Moderate to high (10 ns to 10 μs)	High (<1 ns)	High (<1 ns)	High (<1 ns)
Noise level	Low	Low to moderate	Moderate	Low to moderate
Wavelength range	300–3000 nm (UV to IR)	400–1100 nm (visible to IR)	800–1700 nm (near IR to mid-IR)	200–3000 nm (UV mid-IR)
Quantum efficiency	80%–90%	80%–90%	80%–90%	80%–90%
Fabrication complexity	Moderate	Low	Moderate	High
Cost	Moderate to high	Low	Moderate	High
Flexibility	High	Low	Low	Low
Thermal stability	High	Moderate to high	High	High
Integration capability	High	High (with Si electronics)	Moderate	Moderate
References	[187, 188]	[189, 190]	[191, 192]	[193, 194]

roll-to-roll processing, or printing technologies, making CNT photodetectors more accessible for commercial applications.

CNTs exhibit quantum confinement effects that can be exploited for quantum photonics applications, such as single-photon detection, which is crucial for quantum computing and secure communication. Research in this area might explore how CNTs can be used to develop highly sensitive, low-noise photodetectors for these cutting-edge technologies.

CNT-based photodetectors could play a significant role in environmental monitoring and biomedical applications, such as detecting pollutants or monitoring biological processes in real time. Future research might focus on improving the biocompatibility of CNTs and their integration into portable sensing devices.

Despite the potential, there are challenges such as stability, reproducibility, and the toxicity of CNTs. Future research will likely focus on overcoming these challenges by developing more stable CNT structures, improving fabrication techniques, and conducting thorough safety assessments.

In summary, the future of research on CNT photodetectors is bright, with potential breakthroughs in various fields, including flexible electronics, quantum technologies, and environmental sensing. As the technology matures, CNT-based photodetectors could become a key component in next-generation optoelectronic devices.

4.6 CONCLUSIONS

In conclusion, CNTs represent a groundbreaking advancement in nanotechnology, offering unparalleled mechanical, electrical, thermal, and optical properties that make them highly valuable across a wide range of applications. Their structural variety, from single-walled to multiwalled configurations, provides a foundation for their diverse functionality. The synthesis methods, including arc discharge, laser ablation, and CVD, each contribute unique advantages and challenges in producing high-quality CNTs. The purification processes further refine these materials, ensuring their suitability for practical applications.

CNTs are increasingly utilized in electronics, enhancing the performance of devices and serving as field-emission sources. Their high conductivity and large surface area are leveraged in energy storage technologies, such as batteries and supercapacitors, contributing to advancements in efficient energy solutions. In the medical field, CNTs are being explored for drug delivery systems, gene therapy, and biosensing, indicating their potential to revolutionize healthcare. Their use in gas and hydrogen storage highlights their role in addressing energy challenges, while their incorporation into composite materials demonstrates their ability to improve traditional materials. The application of CNTs in photodetectors, including various types like photoconductors and photodiodes, underscores their versatility and importance in developing advanced sensing technologies.

Despite these significant advancements, several research gaps remain that warrant further exploration. One critical area is the need for improved methods of scaling up CNT synthesis while maintaining high quality and uniformity. Current synthesis techniques often face challenges in producing CNTs with consistent diameters and chiralities, which are crucial for specific applications. Additionally, the development of cost-effective and efficient purification processes is necessary to enhance the practical usability of CNTs in large-scale applications.

Another area requiring attention is the thorough understanding of the long-term stability and environmental impact of CNTs. Research into their biocompatibility, degradation, and potential ecological effects is essential to ensure their safe use in various fields, particularly in medicine and environmental applications. Moreover, while the electrical and mechanical properties of CNTs are well-documented, there is a need for more comprehensive studies on their interaction with other materials in composite structures, as well as their behavior under extreme conditions.

Finally, advancements in photodetector technology using CNTs highlight the need for further research into optimizing performance metrics such as responsivity, quantum efficiency, and detectivity. Addressing these gaps will enhance the development of more sensitive and efficient photodetection systems.

In conclusion, while CNTs have demonstrated immense potential across various domains, addressing these research gaps will be crucial for realizing their full capabilities and ensuring their widespread and sustainable application in technology and industry.

ACKNOWLEDGMENT

This work has been supported by the Science and Engineering Research Board (SERB) Government of India (Grant No. CRG2022/006428-G).

REFERENCES

1. Davis, R. F. (1992). *Diamond Films and Coatings*. Park Ridge: Noyes Publications.
2. Yahya, N., Nor, R. M., Bakar, S. A., Thandavan, T. M., & Rusop, M. (2011). Diamond: synthesis, characterisation and applications. *Carbon and Oxide Nanostructures: Synthesis, Characterisation and Applications*, 195–217.
3. Bernal, J. D. (1924). The structure of graphite. *Proceedings of the Royal Society of London. Series A, Containing Papers of a Mathematical and Physical Character*, *106*(740), 749–773.
4. Kroto, H. W., Heath, J. R., O'Brien, S. C., Curl, R. F., & Smalley, R. E. (1995). C60: Buckminsterfullerene. *Nature [Internet]*, *318*(6042), 1985
5. Iijima, S. (1991). Helical microtubules of graphitic carbon. *Nature*, *354*(6348), 56–58.
6. Rashid, M. H. O., & Ralph, S. F. (2017). Carbon nanotube membranes: synthesis, properties, and future filtration applications. *Nanomaterials*, *7*(5), 99.
7. Harris, P. J. (2001). *Carbon Nanotubes and Related Structures: New Materials for the Twenty-First Century*. Cambridge University Press.
8. Wang, Y. (1992). Photoconductivity of fullerene-doped polymers. *Nature*, *356*(6370), 585–587.
9. Baker, S. N., & Baker, G. A. (2010). Luminescent carbon nanodots: emergent nanolights. *Angewandte Chemie International Edition*, *49*(38), 6726–6744.
10. Baughman, R. H., Cui, C., Zakhidov, A. A., Iqbal, Z., Barisci, J. N., Spinks, G. M., ... & Kertesz, M. (1999). Carbon nanotube actuators. *Science*, *284*(5418), 1340–1344.
11. Nair, R. R., Blake, P., Grigorenko, A. N., Novoselov, K. S., Booth, T. J., Stauber, T., ... & Geim, A. K. (2008). Fine structure constant defines visual transparency of graphene. *Science*, *320*(5881), 1308–1308.

12. Ponomarenko, L. A., Schedin, F., Katsnelson, M. I., Yang, R., Hill, E. W., Novoselov, K. S., & Geim, A. K. (2008). Chaotic Dirac billiard in graphene quantum dots. *Science, 320*(5874), 356–358.
13. Zeng, Q., Wang, S., Yang, L., Wang, Z., Pei, T., Zhang, Z., ... & Xie, S. (2012). Carbon nanotube arrays based high-performance infrared photodetector. *Optical Materials Express*, *2*(6), 839–848.
14. Freitag, M., Martin, Y., Misewich, J. A., Martel, R., & Avouris, P. (2003). Photoconductivity of single carbon nanotubes. *Nano Letters*, *3*(8), 1067–1071.
15. Chen, H., Xi, N., Song, B., Chen, L., Zhao, J., Lai, K. W. C., & Yang, R. (2012). Infrared camera using a single nano-photodetector. *IEEE Sensors Journal*, *13*(3), 949–958.
16. Lu, R., Christianson, C., Weintrub, B., & Wu, J. Z. (2013). High photoresponse in hybrid graphene–carbon nanotube infrared detectors. *CS Applied Materials & Interfaces*, *5*(22), 11703–11707.
17. Lu, R., Christianson, C., Kirkeminde, A., Ren, S., & Wu, J. (2012). Extraordinary photocurrent harvesting at type-II heterojunction interfaces: toward high detectivity carbon nanotube infrared detectors. *Nano Letters*, *12*(12), 6244–6249.
18. Itkis, M. E., Borondics, F., Yu, A., & Haddon, R. C. (2006). Bolometric infrared photoresponse of suspended single-walled carbon nanotube films. *Science*, *312*(5772), 413–416.
19. Thostenson, E. T., Ren, Z., & Chou, T. W. (2001). Advances in the science and technology of carbon nanotubes and their composites: a review. *Composites Science and Technology*, *61*(13), 1899–1912.
20. Pandey, P., & Dahiya, M. (2016). Carbon nanotubes: types, methods of preparation and applications. *Carbon*, *1*(4), 15–21.
21. Rastogi, V., Yadav, P., Bhattacharya, S. S., Mishra, A. K., Verma, N., Verma, A., & Pandit, J. K. (2014). Carbon nanotubes: an emerging drug carrier for targeting cancer cells. *Journal of Drug Delivery*, *2014*(1), 670815.
22. Mintmire, J. W., & White, C. T. (1995). Electronic and structural properties of carbon nanotubes. *Carbon*, *33*(7), 893–902.
23. Odom, T. W., Huang, J. L., Kim, P., & Lieber, C. M. (2000). Structure and electronic properties of carbon nanotubes. *Journal of Physical Chemistry B*, *104*(13), 2794–2809.
24. Yakobson, B. I., & Avouris, P. (2001). Mechanical properties of carbon nanotubes. In *Carbon Nanotubes: Synthesis, Structure, Properties, and Applications*, 287–327. Berlin, Heidelberg: Springer Berlin Heidelberg.
25. Ruoff, R. S., & Lorents, D. C. (1995). Mechanical and thermal properties of carbon nanotubes. *Carbon*, *33*(7), 925–930.
26. Kumari, L., Zhang, T., Du, G. H., Li, W. Z., Wang, Q. W., Datye, A., & Wu, K. H. (2008). Thermal properties of CNT-alumina nanocomposites. *Composites Science and Technology*, *68*(9), 2178–2183.
27. Ibrahim, K. S. (2013). Carbon nanotubes? Properties and applications: a review. *Carbon Letters*, *14*(3), 131–144.
28. Eatemadi, A., Daraee, H., Karimkhanloo, H., Kouhi, M., Zarghami, N., Akbarzadeh, A., ... & Joo, S. W. (2014). Carbon nanotubes: properties,

synthesis, purification, and medical applications. *Nanoscale Research Letters*, *9*, 1–13.
29. Helland, A., Wick, P., Koehler, A., Schmid, K., & Som, C. (2008). Reviewing the environmental and human health knowledge base of carbon nanotubes. *Ciência & Saúde Coletiva*, *13*, 441–452.
30. Baughman, R. H., Zakhidov, A. A., & De Heer, W. A. (2002). Carbon nanotubes—the route toward applications. *Science*, *297*(5582), 787–792.
31. Cao, A., Zhu, H., Zhang, X., Li, X., Ruan, D., Xu, C., ... & Wu, D. (2001). Hydrogen storage of dense-aligned carbon nanotubes. *Chemical Physics Letters*, *342*(5–6), 510–514.
32. Kar, S., Bindal, R. C., Prabhakar, S., Tewari, P. K., Dasgupta, K., & Sathiyamoorthy, D. (2008). Potential of carbon nanotubes in water purification: an approach towards the development of an integrated membrane system. *International Journal of Nuclear Desalination*, *3*(2), 143–150.
33. Tans, S. J., Verschueren, A. R., & Dekker, C. (1998). Room-temperature transistor based on a single carbon nanotube. *Nature*, *393*(6680), 49–52.
34. Venema, L. C., Wildöer, J. W. G., Tuinstra, H. T., Dekker, C., Rinzler, A. G., & Smalley, R. E. (1997). Length control of individual carbon nanotubes by nano structuring with a scanning tunneling microscope. *Applied Physics Letters*, *71*(18), 2629–2631.
35. Postma, H. W. C., de Jonge, M., Yao, Z., & Dekker, C. (2000). Electrical transport through carbon nanotube junctions created by mechanical manipulation. *Physical Review B*, *62*(16), R10653.
36. Park, J. Y., Yaish, Y., Brink, M., Rosenblatt, S., & McEuen, P. L. (2002). Electrical cutting and nicking of carbon nanotubes using an atomic force microscope. *Applied Physics Letters*, *80*(23), 4446–4448.
37. Frackowiak, E., & Beguin, F. (2002). Electrochemical storage of energy in carbon nanotubes and nanostructured carbons. *Carbon*, *40*(10), 1775–1787.
38. Javey, A., Wang, Q., Ural, A., Li, Y., & Dai, H. (2002). Carbon nanotube transistor arrays for multistage complementary logic and ring oscillators. *Nano Letters*, *2*(9), 929–932.
39. Collins, P. G., Arnold, M. S., & Avouris, P. (2001). Engineering carbon nanotubes and nanotube circuits using electrical breakdown. *Science*, *292*(5517), 706–709.
40. Blase, X., Charlier, J. C., De Vita, De Vita, A., Car, R., Redlich, P., Terrones, M., ... & Ajayan, P. M. (1999). Boron-mediated growth of long helicity-selected carbon nanotubes. *Physical Review Letters*, *83*(24), 5078.
41. De Vita, A., Charlier, J. C., Blase, X., & Car, R. (1999). Electronic structure at carbon nanotube tips. *Applied Physics—Section A—Materials Science and Processing*, *68*(3), 283–286.
42. Bonard, J. M., Stöckli, T., Maier, F., de Heer, W. A., Châtelain, A., Salvetat, J. P., & Forró, L. (1998). Field-emission-induced luminescence from carbon nanotubes. *Physical Review Letters*, *81*(7), 1441.
43. Rinzler, A. G., Hafner, J. H., Nikolaev, P., Nordlander, P., Colbert, D. T., Smalley, R. E., ... & Tománek, D. (1995). Unraveling nanotubes: field emission from an atomic wire. *Science*, *269*(5230), 1550–1553.

44. Saito, Y., Hamaguchi, K., Hata, K., Uchida, K., Tasaka, Y., Ikazaki, F., ... & Nishina, Y. (1997). Conical beams from open nanotubes. *Nature, 389*(6651), 554–555.
45. Yue, G. Z., Qiu, Q., Gao, B., Cheng, Y., Zhang, J., Shimoda, H., ... & Zhou, O. (2002). Generation of continuous and pulsed diagnostic imaging x-ray radiation using a carbon-nanotube-based field-emission cathode. *Applied Physics Letters, 81*(2), 355–357.
46. Saito, Y., Uemura, S., & Hamaguchi, K. (1998). Cathode ray tube lighting elements with carbon nanotube field emitters. *Japanese Journal of Applied Physics, 37*(3B), L346.
47. Sugie, H., Tanemura, M., Filip, V., Iwata, K., Takahashi, K., & Okuyama, F. (2001). Carbon nanotubes as electron source in an x-ray tube. *Applied Physics Letters, 78*(17), 2578–2580.
48. Xia, H., Wang, Y., Lin, J., & Lu, L. (2012). Hydrothermal synthesis of MnO_2/CNT nanocomposite with a CNT core/porous MnO_2 sheath hierarchy architecture for supercapacitors. *Nanoscale Research Letters*, 7, 1–10.
49. Evanoff, K., Benson, J., Schauer, M., Kovalenko, I., Lashmore, D., Ready, W. J., & Yushin, G. (2012). Ultra strong silicon-coated carbon nanotube nonwoven fabric as a multifunctional lithium-ion battery anode. *ACS Nano, 6*(11), 9837–9845.
50. Lee, N. S., Chung, D. S., Han, I. T., Kang, J. H., Choi, Y. S., Kim, H. Y., ... & Kim, J. M. (2001). Application of carbon nanotubes to field emission displays. *Diamond and Related Materials, 10*(2), 265–270.
51. Meunier, V., Kephart, J., Roland, C., & Bernholc, J. (2002). Abinitio investigations of lithium diffusion in carbon nanotube systems. *Physical Review Letters, 88*(7), 075506.
52. Yun, Y. S., Yoon, G., Kang, K., & Jin, H. J. (2014). High-performance supercapacitors based on defect-engineered carbon nanotubes. *Carbon, 80*, 246–254.
53. Frackowiak, E., Jurewicz, K., Delpeux, S., & Béguin, F. (2001). Nanotubular materials for supercapacitors. *Journal of Power Sources, 97*, 822–825.
54. Futaba, D. N., Hata, K., Yamada, T., Hiraoka, T., Hayamizu, Y., Kakudate, Y., ... & Iijima, S. (2006). Shape-engineerable and highly densely packed single-walled carbon nanotubes and their application as supercapacitor electrodes. *Nature Materials, 5*(12), 987–994.
55. An, K. H., Kim, W. S., Park, Y. S., Moon, J. M., Bae, D. J., Lim, S. C., ... & Lee, Y. H. (2001). Electrochemical properties of high-power supercapacitors using single-walled carbon nanotube electrodes. *Advanced Functional Materials, 11*(5), 387–392.
56. Du, C., Yeh, J., & Pan, N. (2005). High power density supercapacitors using locally aligned carbon nanotube electrodes. *Nanotechnology, 16*(4), 350.
57. Lim, S. H., Elim, H. I., Gao, X. Y., Wee, A. T. S., Ji, W., Lee, J. Y., & Lin, J. (2006). Electronic and optical properties of nitrogen-doped multiwalled carbon nanotubes. *Physical Review B—Condensed Matter and Materials Physics, 73*(4), 045402.

58. Sun, J., Huang, Y., Fu, C., Wang, Z., Huang, Y., Zhu, M., ... & Hu, H. (2016). High-performance stretchable yarn supercapacitor based on PPy@ CNTs@ urethane elastic fiber core spun yarn. *Nano Energy*, *27*, 230–237.
59. Song, C., Yun, J., Keum, K., Jeong, Y. R., Park, H., Lee, H., ... & Ha, J. S. (2019). High performance wire-type supercapacitor with Ppy/CNT-ionic liquid/AuNP/carbon fiber electrode and ionic liquid based electrolyte. *Carbon*, *144*, 639–648.
60. Elhissi, A., Ahmed, W., Dhanak, V., & Subramani, K. (2012). Carbon nanotubes in cancer therapy and drug delivery. In *Emerging Nanotechnologies in Dentistry*, 347–363. William Andrew Publishing.
61. Wong, B. S., Yoong, S. L., Jagusiak, A., Panczyk, T., Ho, H. K., Ang, W. H., & Pastorin, G. (2013). Carbon nanotubes for delivery of small molecule drugs. *Advanced Drug Delivery Reviews*, *65*(15), 1964–2015.
62. Podesta, J. E., Al-Jamal, K. T., Herrero, M. A., Tian, B., Ali-Boucetta, H., Hegde, V., ... & Kostarelos, K. (2009). Antitumor activity and prolonged survival by carbon-nanotube-mediated therapeutic siRNA silencing in a human lung xenograft model. *Small*, *5*(10), 1176–1185.
63. Gooding, J. J. (2005). Nanostructuring electrodes with carbon nanotubes: a review on electrochemistry and applications for sensing. *Electrochimica Acta*, *50*(15), 3049–3060.
64. Zhao, Y. D., Zhang, W. D., Chen, H., & Luo, Q. M. (2002). Direct electron transfer of glucose oxidase molecules adsorbed onto carbon nanotube powder microelectrode. *Analytical Sciences*, *18*(8), 939–941.
65. Munge, B., Liu, G., Collins, G., & Wang, J. (2005). Multiple enzyme layers on carbon nanotubes for electrochemical detection down to 80 DNA copies. *Analytical Chemistry*, *77*(14), 4662–4666.
66. Banhart, F., Grobert, N., Terrones, M., Charlier, J. C., & Ajayan, P. M. (2001). Metal atoms in carbon nanotubes and related nanoparticles. *International Journal of Modern Physics B*, *15*(31), 4037–4069.
67. Park, S. J., & Lee, S. Y. (2009). Hydrogen storage behaviors of carbon nanotubes/metal-organic frameworks-5 hybrid composites. *Carbon Letters*, *10*(1), 19–22.
68. Liu, C., Fan, Y. Y., Liu, M., Cong, H. T., Cheng, H. M., & Dresselhaus, M. S. (1999). Hydrogen storage in single-walled carbon nanotubes at room temperature. *Science*, *286*(5442), 1127–1129.
69. Gadd, G. E., Blackford, M., Moricca, S., Webb, N., Evans, P. J., Smith, A. M., ... & Hua, Q. (1997). The world's smallest gas cylinders? *Science*, *277*(5328), 933–936.
70. Terrones, M., Kamalakaran, R., Seeger, T., & Rühle, A. M. (2000). Novel nanoscale gas containers: encapsulation of N_2 in CNx nanotubes. *Chemical Communications*, (23), 2335–2336.
71. Trasobares, S., Stephan, O., Colliex, C., Hug, G., Hsu, W. K., Kroto, H. W., & Walton, D. R. M. (2001). Electron beam puncturing of carbon nanotube containers for release of stored N_2 gas. *European Physical Journal B—Condensed Matter and Complex Systems*, *22*, 117–122.
72. Chambers, A., Park, C., Baker, R. T. K., & Rodriguez, N. M. (1998). Hydrogen storage in graphite nanofibers. *Journal of Physical Chemistry B*, *102*(22), 4253–4256.

73. Chen, P., Wu, X., Lin, J., & Tan, K. L. (1999). High H_2 uptake by alkali-doped carbon nanotubes under ambient pressure and moderate temperatures. *Science*, *285*(5424), 91–93.
74. Hirscher, M., Becher, M., Haluska, M., Quintel, A., Skakalova, V., Choi, Y. M., ... & Fink, J. (2002). Hydrogen storage in carbon nanostructures. *Journal of Alloys and Compounds*, *330*, 654–658.
75. Gordon, P. A., & Saeger, R. B. (1999). Molecular modeling of adsorptive energy storage: hydrogen storage in single-walled carbon nanotubes. *Industrial & Engineering Chemistry Research*, *38*(12), 4647–4655.
76. Meregalli, V., & Parrinello, M. (2001). Review of theoretical calculations of hydrogen storage in carbon-based materials. *Applied Physics A*, *72*, 143–146.
77. Lee, S. M., An, K. H., Lee, Y. H., Seifert, G., & Frauenheim, T. (2001). A hydrogen storage mechanism in single-walled carbon nanotubes. *Journal of the American Chemical Society*, *123*(21), 5059–5063.
78. Darkrim, F. L., Malbrunot, P., & Tartaglia, G. P. (2002). Review of hydrogen storage by adsorption in carbon nanotubes. *International Journal of Hydrogen Energy*, *27*(2), 193–202.
79. Tanaka, H., El-Merraoui, M., Steele, W. A., & Kaneko, K. (2002). Methane adsorption on single-walled carbon nanotube: a density functional theory model. *Chemical Physics Letters*, *352*(5–6), 334–341.
80. Meyyappan, M. (2004). *Carbon Nanotubes: Science and Applications*. CRC Press.
81. Ounaies, Z., Park, C., Wise, K. E., Siochi, E. J., & Harrison, J. S. (2003). Electrical properties of single wall carbon nanotube reinforced polyimide composites. *Composites Science and Technology*, *63*(11), 1637–1646.
82. Dresselhaus, M. S. (2001). Burn and interrogate. *Science*, *292*(5517), 650–651.
83. Lee, J. U. (2005). Photovoltaic effect in ideal carbon nanotube diodes. *Applied Physics Letters*, *87*(7).
84. Ago, H., Petritsch, K., Shaffer, M. S., Windle, A. H., & Friend, R. H. (1999). Composites of carbon nanotubes and conjugated polymers for photovoltaic devices. *Advanced Materials*, *11*(15), 1281–1285.
85. Kymakis, E., Kornilios, N., & Koudoumas, E. (2008). Carbon nanotube doping of P3HT: PCBM photovoltaic devices. *Journal of Physics D: Applied Physics*, *41*(16), 165110.
86. Maheswaran, R., & Shanmugavel, B. P. (2022). A critical review of the role of carbon nanotubes in the progress of next-generation electronic applications. *Journal of Electronic Materials*, *51*(6), 2786–2800.
87. Lee, J., & Kim, S. (2005). Manufacture of a nanotweezer using a length controlled CNT arm. *Sensors and Actuators A: Physical*, *120*(1), 193–198.
88. Someya, T., Small, J., Kim, P., Nuckolls, C., & Yardley, J. T. (2003). Alcohol vapor sensors based on single-walled carbon nanotube field effect transistors. *Nano Letters*, *3*(7), 877–881.
89. Mahar, B., Laslau, C., Yip, R., & Sun, Y. (2007). Development of carbon nanotube-based sensors—a review. *IEEE Sensors Journal*, *7*(2), 266–284.

90. Kawano, T., Chiamori, H. C., Suter, M., Zhou, Q., Sosnowchik, B. D., & Lin, L. (2007). An electrothermal carbon nanotube gas sensor. *Nano Letters*, *7*(12), 3686–3690.
91. Paradise, M., & Goswami, T. (2007). Carbon nanotubes—production and industrial applications. *Materials & Design*, *28*(5), 1477–1489.
92. Avouris, P., Freitag, M., & Perebeinos, V. (2008). Carbon-nanotube photonics and optoelectronics. *Nature Photonics*, *2*(6), 341–350.
93. Bonaccorso, F., Sun, Z., Hasan, T., & Ferrari, A. C. (2010). Graphene photonics and optoelectronics. *Nature Photonics*, *4*(9), 611–622.
94. Meunier, V., Souza Filho, A. G., Barros, E. B., & Dresselhaus, M. S. (2016). Physical properties of low-dimensional SP2-based carbon nanostructures. *Reviews of Modern Physics*, *88*(2), 025005.
95. Itkis, M. E., Niyogi, S., Meng, M. E., Hamon, M. A., Hu, H., & Haddon, R. C. (2002). Spectroscopic study of the Fermi level electronic structure of single-walled carbon nanotubes. *Nano Letters*, *2*(2), 155–159.
96. Buldum, A., & Lu, J. P. (2001). Contact resistance between carbon nanotubes. *Physical Review B*, *63*(16), 161403.
97. He, X., Léonard, F., & Kono, J. (2015). Uncooled carbon nanotube photodetectors. *Advanced Optical Materials*, *3*(8), 989–1011.
98. Nirmalraj, P. N., Lyons, P. E., De, S., Coleman, J. N., & Boland, J. J. (2009). Electrical connectivity in single-walled carbon nanotube networks. *Nano Letters*, *9*(11), 3890–3895.
99. Lyons, P. E., De, S., Blighe, F., Nicolosi, V., Pereira, L. F. C., Ferreira, M. S., & Coleman, J. N. (2008). The relationship between network morphology and conductivity in nanotube films. *Journal of Applied Physics*, *104*(4).
100. Snow, E. S., Novak, J. P., Campbell, P. M., & Park, D. (2003). Random networks of carbon nanotubes as an electronic material. *Applied Physics Letters*, *82*(13), 2145–2147.
101. Xia, F., Mueller, T., Lin, Y. M., Valdes-Garcia, A., & Avouris, P. (2009). Ultrafast graphene photodetector. *Nature Nanotechnology*, *4*(12), 839–843.
102. Koppens, F. H. L., Mueller, T., Avouris, P., Ferrari, A. C., Vitiello, M. S., & Polini, M. (2014). Photodetectors based on graphene, other two-dimensional materials and hybrid systems. *Nature Nanotechnology*, *9*(10), 780–793.
103. Lv, R., Cruz-Silva, E., & Terrones, M. (2014). Building complex hybrid carbon architectures by covalent interconnections: graphene–nanotube hybrids and more. *ACS Nano*, *8*(5), 4061–4069.
104. Liu, Y., Liu, Y., Qin, S., Xu, Y., Zhang, R., & Wang, F. (2017). Graphene-carbon nanotube hybrid films for high-performance flexible photodetectors. *Nano Research*, *10*, 1880–1887.
105. Cao, X., Liu, X., Li, X., Lei, X., & Chen, W. (2018). Conductive stability of graphene on PET and glass substrates under blue light irradiation. *Optics Communications*, *406*, 169–172.
106. Lin, X., Liu, P., Wei, Y., Li, Q., Wang, J., Wu, Y., ... & Jiang, K. (2013). Development of an ultra-thin film consisting of a graphene membrane and carbon nanotube vein support. *Nature Communications*, *4*(1), 2920.

107. Liu, Y., Wang, F., Wang, X., Wang, X., Flahaut, E., Liu, X., ... & Zhang, R. (2015). Planar carbon nanotube–graphene hybrid films for high-performance broadband photodetectors. *Nature Communications*, *6*(1), 8589.
108. Shi, J., Li, X., Cheng, H., Liu, Z., Zhao, L., Yang, T., ... & Fang, Y. (2016). Graphene reinforced carbon nanotube networks for wearable strain sensors. *Advanced Functional Materials*, *26*(13), 2078–2084.
109. Konstantatos, G., Badioli, M., Gaudreau, L., Osmond, J., Bernechea, M., De Arquer, F. P. G., ... & Koppens, F. H. (2012). Hybrid graphene–quantum dot phototransistors with ultrahigh gain. *Nature Nanotechnology*, *7*(6), 363–368.
110. Liu, Y., Wang, F., Liu, Y., Wang, X., Xu, Y., & Zhang, R. (2016). Charge transfer at carbon nanotube–graphene van der Waals heterojunctions. *Nanoscale*, *8*(26), 12883–12886.
111. Huang, L., Huang, Y., Liang, J., Wan, X., & Chen, Y. (2011). Graphene-based conducting inks for direct inkjet printing of flexible conductive patterns and their applications in electric circuits and chemical sensors. *Nano Research*, *4*, 675–684.
112. Akinwande, D., Petrone, N., & Hone, J. (2014). Two-dimensional flexible nanoelectronics. *Nature Communications*, *5*(1), 5678.
113. Wang, K., Wu, H., Meng, Y., Zhang, Y., & Wei, Z. (2012). Integrated energy storage and electrochromic function in one flexible device: an energy storage smart window. *Energy & Environmental Science*, *5*(8), 8384–8389.
114. Yeh, M. H., Lin, L., Yang, P. K., & Wang, Z. L. (2015). Motion-driven electrochromic reactions for self-powered smart window systems. *ACS Nano*, *9*(5), 4757–4765.
115. Kim, D. H., Lu, N., Ma, R., Kim, Y. S., Kim, R. H., Wang, S., ... & Rogers, J. A. (2011). Epidermal electronics. *Science*, *333*(6044), 838–843.
116. Liang, Y., Wang, Z., Huang, J., Cheng, H., Zhao, F., Hu, Y., ... & Qu, L. (2015). Series of in-fiber graphene supercapacitors for flexible wearable devices. *Journal of Materials Chemistry A*, *3*(6), 2547–2551.
117. Nathan, A., Ahnood, A., Cole, M. T., Lee, S., Suzuki, Y., Hiralal, P., ... & Milne, W. I. (2012). Flexible electronics: the next ubiquitous platform. *Proceedings of the IEEE, 100 (Special Centennial Issue)*, 1486–1517.
118. Kholmanov, I. N., Magnuson, C. W., Piner, R., Kim, J. Y., Aliev, A. E., Tan, C., ... & Ruoff, R. S. (2015). Optical, electrical, and electromechanical properties of hybrid graphene/carbon nanotube films. *Advanced Materials*, *27*(19), 3053–3059.
119. Xiao, L., Ma, H., Liu, J., Zhao, W., Jia, Y., Zhao, Q., ... & Jiang, K. (2015). Fast adaptive thermal camouflage based on flexible VO_2/graphene/CNT thin films. *Nano Letters*, *15*(12), 8365–8370.
120. Pradhan, B., Setyowati, K., Liu, H., Waldeck, D. H., & Chen, J. (2008). Carbon nanotube—polymer nanocomposite infrared sensor. *Nano Letters*, *8*(4), 1142–1146.
121. Levitsky, I. A., & Euler, W. B. (2003). Photoconductivity of single-wall carbon nanotubes under continuous-wave near-infrared illumination. *Applied Physics Letters*, *83*(9), 1857–1859.

122. Fujiwara, A., Matsuoka, Y., Suematsu, H., Ogawa, N., Miyano, K., Kataura, H., ... & Achiba, Y. (2001, October). Photoconductivity of single-walled carbon nanotubes. *AIP Conference Proceedings*, *590*(1), 189–192.
123. Park, S., Kim, S. J., Nam, J. H., Pitner, G., Lee, T. H., Ayzner, A. L., ... & Bao, Z. (2014). Significant enhancement of infrared photodetector sensitivity using a semiconducting single-walled carbon nanotube/C60 phototransistor. *Advanced Materials (Deerfield Beach, Fla.)*, *27*(4), 759–765.
124. Cai, X., Wang, S., & Peng, L. M. (2023). Recent progress of photodetector based on carbon nanotube film and application in optoelectronic integration. *Nano Research Energy*, *2*(2), e9120058.
125. Lu, R., Li, Z., Xu, G., & Wu, J. Z. (2009). Suspending single-wall carbon nanotube thin film infrared bolometers on microchannels. *Applied Physics Letters*, *94*(16).
126. Zhang, Q., Yan, R., Peng, X., Wang, Y., & Feng, S. (2022). TiO_{2-x} films for bolometer applications: recent progress and perspectives. *Materials Research Express*, *9*(1), 012002.
127. Tarasov, M., Svensson, J., Kuzmin, L., & Campbell, E. E. (2007). Carbon nanotube bolometers. *Applied Physics Letters*, *90*(16).
128. Fernandes, G. E., Kim, J. H., Sood, A. K., & Xu, J. (2013). Giant temperature coefficient of resistance in carbon nanotube/phase-change polymer nanocomposites. *Advanced Functional Materials*, *23*(37), 4678–4683.
129. Pradhan, B., Kohlmeyer, R. R., Setyowati, K., Owen, H. A., & Chen, J. (2009). Advanced carbon nanotube/polymer composite infrared sensors. *Carbon*, *47*(7), 1686–1692.
130. Bernardi, M., Lohrman, J., Kumar, P. V., Kirkeminde, A., Ferralis, N., Grossman, J. C., & Ren, S. (2012). Nanocarbon-based photovoltaics. *ACS Nano*, *6*(10), 8896–8903.
131. Xie, Y., Gong, M., Shastry, T. A., Lohrman, J., Hersam, M. C., & Ren, S. (2013). Broad-spectral-response nanocarbon bulk-heterojunction excitonic photodetectors. *Advanced Materials (Deerfield Beach, Fla.)*, *25*(25), 3433.
132. Long, R., & Prezhdo, O. V. (2014). Asymmetry in the electron and hole transfer at a polymer–carbon nanotube heterojunction. *Nano Letters*, *14*(6), 3335–3341.
133. Mohite, A. D., Gopinath, P., Shah, H. M., & Alphenaar, B. W. (2008). Exciton dissociation and stark effect in the carbon nanotube photocurrent spectrum. *Nano Letters*, *8*(1), 142–146.
134. Mohite, A., Chakraborty, S., Gopinath, P., Sumanasekera, G. U., & Alphenaar, B. W. (2005). Displacement current detection of photoconduction in carbon nanotubes. *Applied Physics Letters*, *86*(6).
135. Fernandes, G. E., Ho Kim, J., Chin, M., Dhar, N., & Xu, J. (2014). Carbon nanotube microbolometers on suspended silicon nitride via vertical fabrication procedure. *Applied Physics Letters*, *104*(20).
136. John, J., Muthee, M., Yogeesh, M., Yngvesson, S. K., & Carter, K. R. (2014). Suspended multiwall carbon nanotube-based infrared sensors via roll-to-roll fabrication. *Advanced Optical Materials*, *2*(6), 581–587.

137. Mahjouri-Samani, M., Zhou, Y. S., He, X. N., Xiong, W., Hilger, P., & Lu, Y. F. (2012). Plasmonic-enhanced carbon nanotube infrared bolometers. *Nanotechnology*, *24*(3), 035502.
138. Vera-Reveles, G., Simmons, T. J., Bravo-Sánchez, M., Vidal, M. A., Navarro-Contreras, H., & González, F. J. (2011). High-sensitivity bolometers from self-oriented single-walled carbon nanotube composites. *ACS Applied Materials & Interfaces*, *3*(8), 3200–3204.
139. Glamazda, A. Y., Karachevtsev, V. A., Euler, W. B., & Levitsky, I. A. (2012). Achieving high mid-IR bolometric responsivity for anisotropic composite materials from carbon nanotubes and polymers. *Advanced Functional Materials*, *22*(10), 2177–2186.
140. Aliev, A. E. (2008). Bolometric detector on the basis of single-wall carbon nanotube/polymer composite. *Infrared Physics & Technology*, *51*(6), 541–545.
141. Bang, D., Lee, J., Park, J., Choi, J., Chang, Y. W., Yoo, K. H., ... & Haam, S. (2012). Effectively enhanced sensitivity of a polyaniline–carbon nanotube composite thin film bolometric near-infrared sensor. *Journal of Materials Chemistry*, *22*(7), 3215–3219.
142. Riedel, W., Fernandes, G. E., Kim, J. H., & Xu, J. (2012). TCR, Responsivity, and Noise Properties of Carbon Nanotube-polymer Bolometric IR Detectors. In *SENSOR DEVICES 2012: The Third International Conference on Sensor Device Technologies and Applications*, Editor: Yurish, S. International Academy, Research, and Industry Association (IARIA).
143. Lu, R., Shi, J. J., Baca, F. J., & Wu, J. Z. (2010). High performance multiwall carbon nanotube bolometers. *Journal of Applied Physics*, *108*(8).
144. Xiao, L., Zhang, Y., Wang, Y., Liu, K., Wang, Z., Li, T., ... & Jiang, K. (2010). A polarized infrared thermal detector made from super-aligned multiwalled carbon nanotube films. *Nanotechnology*, *22*(2), 025502.
145. Mérel, P., Kpetsu, J. B. A., Koechlin, C., Maine, S., Haidar, R., Pélouard, J. L., ... & Paradis, S. (2010). Infrared sensors based on multi-wall carbon nanotube films. *Comptes Rendus. Physique*, *11*(5–6), 375–380.
146. Gohier, A., Dhar, A., Gorintin, L., Bondavalli, P., Bonnassieux, Y., & Cojocaru, C. S. (2011). All-printed infrared sensor based on multiwalled carbon nanotubes. *Applied Physics Letters*, *98*(6).
147. Afrin, R., Khaliq, J., Islam, M., Gul, I. H., Bhatti, A. S., & Manzoor, U. (2012). Synthesis of multiwalled carbon nanotube-based infrared radiation detector. *Sensors and Actuators A: Physical*, *187*, 73–78.
148. Narita, K., Kuribayashi, R., Altintas, E., Someya, H., Tsuda, K., Ohashi, K., ... & Komatsubara, S. (2013). A plastic-based bolometer array sensor using carbon nanotubes for low-cost infrared imaging devices. *Sensors and Actuators A: Physical*, *195*, 142–147.
149. Kozlov, M. E. (2013). Junction array carbon nanotube bolometer. *Journal of Applied Physics*, *113*(16).
150. Niklaus, F., Vieider, C., & Jakobsen, H. (2008). MEMS-based uncooled infrared bolometer arrays: a review. *MEMS/MOEMS Technologies and Applications III*, *6836*, 125–139.

151. He, X., Fujimura, N., Lloyd, J. M., Erickson, K. J., Talin, A. A., Zhang, Q., … & Kono, J. (2014). Carbon nanotube terahertz detector. *Nano Letters*, *14*(7), 3953–3958.
152. St-Antoine, B. C., Ménard, D., & Martel, R. (2012). Photothermoelectric effects in single-walled carbon nanotube films: Reinterpreting scanning photocurrent experiments. *Nano Research*, *5*, 73–81.
153. St-Antoine, B. C., Ménard, D., & Martel, R. (2009). Position sensitive photothermoelectric effect in suspended single-walled carbon nanotube films. *Nano Letters*, *9*(10), 3503–3508.
154. Dillner, U., Kessler, E., & Meyer, H. G. (2013). Figures of merit of thermoelectric and bolometric thermal radiation sensors. *Journal of Sensors and Sensor Systems*, *2*(1), 85–94.
155. Hu, C., Liu, C., Chen, L., Meng, C., & Fan, S. (2010). A demo opto-electronic power source based on single-walled carbon nanotube sheets. *ACS Nano*, *4*(8), 4701–4706.
156. St-Antoine, B. C., Ménard, D., & Martel, R. (2011). Single-walled carbon nanotube thermopile for broadband light detection. *Nano Letters*, *11*(2), 609–613.
157. Omari, M., & Kouklin, N. A. (2011). Photothermovoltaic effect in carbon nanotubes: en route toward junctionless infrared photocells and light sensors. *Applied Physics Letters*, *98*(24).
158. He, X., Wang, X., Nanot, S., Cong, K., Jiang, Q., Kane, A. A., … & Kono, J. (2013). Photothermoelectric *p–n* junction photodetector with intrinsic broadband polarimetry based on macroscopic carbon nanotube films. *ACS Nano*, *7*(8), 7271–7277.
159. Nanot, S., Cummings, A. W., Pint, C. L., Ikeuchi, A., Akiho, T., Sueoka, K., … & Kono, J. (2013). Broadband, polarization-sensitive photodetector based on optically-thick films of macroscopically long, dense and aligned carbon nanotubes. *Scientific Reports*, *3*(1), 1335.
160. Tseng, Y. H., He, Y., & Que, L. (2013). Ultrasensitive thin film infrared sensors enabled by hybrid nanomaterials. *Analyst*, *138*(10), 3053–3057.
161. Fedorov, G., Kardakova, A., Gayduchenko, I., Charayev, I., Voronov, B. M., Finkel, M., Klapwijk, T. M., Morozov, S., Presniakov, M., Bobrinetskiy, I., Ibragimov, R., Goltsman, G. (2013). Photothermoelectric response in asymmetric carbon nanotube devices exposed to sub-terahertz radiation. *Applied Physics Letters*, *103*(585), 181121.
162. Wang, Y., Zhang, G., Qiao, L., Wei, J., Zhu, J. L., Chen, Z., … & Sun, J. L. (2014). Photocurrent response of carbon nanotube–metal heterojunctions in the terahertz range. *Optics Express*, *22*(5), 5895–5903.
163. Wang, F., Dukovic, G., Brus, L. E., & Heinz, T. F. (2005). The optical resonances in carbon nanotubes arise from excitons. *Science*, *308*(5723), 838–841.
164. Maultzsch, J., Pomraenke, R., Reich, S., Chang, E., Prezzi, D., Ruini, A., … & Lienau, C. (2005). Exciton binding energies in carbon nanotubes from two-photon photoluminescence. *Physical Review B—Condensed Matter and Materials Physics*, *72*(24), 241402.
165. Rao, F., Liu, X., Li, T., Zhou, Y., & Wang, Y. (2009). The synthesis and fabrication of horizontally aligned single-walled carbon nanotubes suspended

across wide trenches for infrared detecting application. *Nanotechnology*, *20*(5), 055501.

166. Bindl, D. J., Wu, M. Y., Prehn, F. C., & Arnold, M. S. (2011). Efficiently harvesting excitons from electronic type-controlled semiconducting carbon nanotube films. *Nano Letters*, *11*(2), 455–460.
167. Yang, L., Wang, S., Zeng, Q., Zhang, Z., & Peng, L. M. (2013). Carbon nanotube photoelectronic and photovoltaic devices and their applications in infrared detection. *Small*, *9*(8), 1225–1236.
168. Zeng, Q., Wang, S., Yang, L., Wang, Z., Zhang, Z., Peng, L., ... & Xie, S. (2012). Doping-free fabrication of carbon nanotube thin-film diodes and their photovoltaic characteristics. *Nano Research*, *5*, 33–42.
169. Xie, Y., Lohrman, J., & Ren, S. (2014). Phase aggregation and morphology effects on nanocarbon optoelectronics. *Nanotechnology*, *25*(48), 485601.
170. Arnold, M. S., Zimmerman, J. D., Renshaw, C. K., Xu, X., Lunt, R. R., Austin, C. M., & Forrest, S. R. (2009). Broad spectral response using carbon nanotube/organic semiconductor/C60 photodetectors. *Nano Letters*, *9*(9), 3354–3358.
171. Bindl, D. J., Safron, N. S., & Arnold, M. S. (2010). Dissociating excitons photogenerated in semiconducting carbon nanotubes at polymeric photovoltaic heterojunction interfaces. *ACS Nano*, *4*(10), 5657–5664.
172. Jain, R. M., Howden, R., Tvrdy, K., Shimizu, S., Hilmer, A. J., McNicholas, T. P., ... & Strano, M. S. (2012). Polymer-free near-infrared photovoltaics with single chirality (6, 5) semiconducting carbon nanotube active layers. *Advanced Materials (Deerfield Beach, FL)*, *24*(32), 4436–4439.
173. Bindl, D. J., & Arnold, M. S. (2013). Efficient exciton relaxation and charge generation in nearly monochiral (7, 5) carbon nanotube/C60 thin-film photovoltaics. *Journal of Physical Chemistry C*, *117*(5), 2390–2395.
174. Biswas, C., Jeong, H., Jeong, M. S., Yu, W. J., Pribat, D., & Lee, Y. H. (2013). Quantum dot–carbon nanotube hybrid phototransistor with an enhanced optical stark effect. *Advanced Functional Materials*, *23*(29), 3653–3660.
175. Decoster, D., Harari, J. (2013). *Optoelectronic Sensors*. John Wiley & Sons.
176. He, Z. B., Maurice, J. L., Lee, C. S., Cojocaru, C. S., & Pribat, D. (2010). Nickel catalyst faceting in plasma-enhanced direct current chemical vapor deposition of carbon nanofibers. *Arabian Journal for Science and Engineering*, *35*(1C), 19.
177. Varshney, D., Weiner, B. R., & Morell, G. (2010). Growth and field emission study of a monolithic carbon nanotube/diamond composite. *Carbon*, *48*(12), 3353–3358.
178. Zhang, M., & Li, J. (2009). Carbon nanotube in different shapes. *Materials Today*, *12*(6), 12–18.
179. Obitayo, W., & Liu, T. (2012). A review: carbon nanotube-based piezoresistive strain sensors. *Journal of Sensors*, *2012*(1), 652438.
180. The transmission electron microscope (TEM) images of a MWNT. 2014. www.researchgate.net/figure/Transmission-electron-microscopy-TEM-images-of-multiwalled-carbon-nanotubes-MWCNT-a_fig2_331295635.

181. Liu, X., Lee, C., Zhou, C., & Han, J. (2001). Carbon nanotube field-effect inverters. *Applied Physics Letters*, *79*(20), 3329–3331.
182. Chu, L. L., Que, L., & Gianchandani, Y. B. (2002). Measurements of material properties using differential capacitive strain sensors. *Journal of Microelectromechanical Systems*, *11*(5), 489–498.
183. Wang, K., et al. (2010). Carbon nanotube photodetectors with high sensitivity and ultrafast response. *Nano Letters*, *10*(10), 3342–3348.
184. Wang, X. et al. (2011). High-performance photodetectors based on carbon nanotubes with enhanced sensitivity. *Applied Physics Letters*, *98*(16), 161104.
185. Lee, H. J., et al. (2008). Silicon photodetectors for optical communication systems. *IEEE Journal of Selected Topics in Quantum Electronics*, *14*(6), 1590–1596.
186. Cross, G. H., et al. (2006). Silicon photodetectors and their applications. *IEEE Journal of Quantum Electronics*, *42*(8), 794–803.
187. Bergemann, K., & Leonard, F. (2020). Giga-gain at room temperature in functionalized carbon nanotube phototransistors based on a nonequilibrium mechanism. *ACS Nano*, *14*(8), 10421–10427.
188. Chang, A. K. K., et al. (2012). InGaAs photodetectors for near infrared applications. *Applied Physics Letters*, *100*(19), 191102.
189. Zhang et al. (2014). High-speed InGaAs photodetectors for communication systems. *IEEE Journal of Selected Topics in Quantum Electronics*, *20*(6), 1280–1286.
190. Hughes, M. A.Homewood, K. P., Curry, R. J., Ohno, Y., & Mizutani, T. J. A. P. L. (2013). An ultra-low leakage current single carbon nanotube diode with split-gate and asymmetric contact geometry. *Applied Physics Letters*, *103*(13).
191. Xia, X., Zhou, S., Wang, Y., & Zhang, Z. (2025). Emerging optoelectronic architectures in carbon nanotube photodetector technologies. *Fundamental Research*, *5*(3), 1153–1168.
192. Hsu, C. Y., Lien, D. H., Lu, S. Y., Chen, C. Y., Kang, C. F., Chueh, Y. L., ... & He, J. H. (2012). Supersensitive, ultrafast, and broad-band light-harvesting scheme employing carbon nanotube/TiO_2 core–shell nanowire geometry. *ACS Nano*, 6(8), 6687–6692.
193. Adinolfi, V., & Sargent, E. H. (2017). Photovoltage field-effect transistors. *Nature*, *542*(7641), 324–327.
194. Zhu, Q. B., Li, B., Yang, D. D., Liu, C., Feng, S., Chen, M. L., ... & Sun, D. M. (2021). A flexible ultrasensitive optoelectronic sensor array for neuromorphic vision systems. *Nature Communications*, *12*(1), 1798.
195 *Bindl, D. J., Wu, M. Y ., Prehn, F. C., & Arnold, M. S.* (2011). Efficiently harvesting excitons from electronic type-controlled semiconducting carbon nanotube films. *Nano Letters*, *11*(2), 455–460.
196 Park, S., Kim, S. J., Nam, J. H., Pitner, G., Lee, T. H., Ayzner, A. L., ... & Bao, Z. (2014). Significant enhancement of infrared photodetector sensitivity using a semiconducting single-walled carbon nanotube/C60 phototransistor. *Advanced Materials (Deerfield Beach, Fla.)*, *27*(4), 759–765.

Chapter 5

Graphene and device applications

Roshini K. Thumpakara

5.1 INTRODUCTION

Graphene is a ground-breaking material consisting of a single layer of carbon atoms arranged in a two-dimensional (2D) honeycomb lattice. Its discovery has marked a significant milestone in the field of material science. Often hailed as a miracle material, graphene holds immense potential across various technological applications, from electronics to energy storage and biomedical devices. Graphene, although conceptually understood for decades, was first isolated and characterized in 2004 by two physicists, Andre Geim and Konstantin Novoselov, at the University of Manchester. Prior to their discovery, graphene had been theorized to exist as a fundamental building block of graphite, but it was thought to be unstable in its free-standing form. Graphite, a material used in pencils, is composed of millions of graphene layers stacked together, and each of these layers could theoretically exhibit unique properties. However, isolating a single layer of carbon atoms was considered impossible at the time. Geim and Novoselov's breakthrough came through a relatively simple but ingenious technique known as mechanical exfoliation or the Scotch tape method. Using adhesive tape, they peeled off thin layers of graphite and continued thinning it down until they achieved a single layer of graphene. Once isolated, they confirmed the material's existence using atomic force microscopy (AFM) and various other characterization techniques. This method, although not suited for large-scale production, allowed them to successfully isolate and study the material's extraordinary properties for the first time. Their discovery earned them the Nobel Prize in Physics *in 2010*, as it opened the door to the exploration of a wide range of 2D materials and spurred a new era in material science.

DOI: 10.1201/9781003512899-5

5.2 ATOMIC STRUCTURE OF GRAPHENE

Graphene's atomic structure is a critical factor in understanding its properties. It consists of carbon atoms arranged in a honeycomb-like pattern, where each carbon atom is bonded to three others via sp2 hybridization.

This results in the formation of strong sigma (σ) bonds in the plane of the lattice, while the remaining electron forms a delocalized pi (π) bond, contributing to graphene's extraordinary electrical conductivity. These σ-bonds are responsible for the remarkable mechanical strength of graphene. The bond length between carbon atoms in graphene is approximately 0.142 nm, and the hexagonal arrangement ensures a uniform, stable structure. Graphene's 2D structure gives it several unique properties, including its exceptional strength, lightweight nature, flexibility, and superior electrical and thermal conductivity. Despite being just one atom thick, graphene is incredibly robust, making it one of the strongest known materials.

5.3 METHODS OF GRAPHENE SYNTHESIS

Several techniques exist to synthesize graphene, with each method offering different advantages and limitations depending on the desired quality and scale of production. The primary methods include the following.

5.3.1 Mechanical exfoliation

This method, also known as the "Scotch tape" technique, was used to first isolate graphene from graphite. While it produces high-quality graphene, it is not suitable for large-scale production.

5.3.2 Chemical vapor deposition

It is one of the most widely used methods for synthesizing high-quality graphene on a large scale. In this process, a hydrocarbon gas (such as methane or acetylene) is introduced into a reaction chamber containing a heated metal substrate, typically copper or nickel, which serves as a catalyst for graphene growth. Chemical vapor deposition (CVD) allows for the production of continuous, large-area graphene sheets suitable for various applications. It is scalable for industrial production, making it a preferred method for commercial applications. But after synthesis, transferring graphene from the metal substrate to other surfaces is challenging and can introduce defects. Also, the process requires precise temperature control and expensive equipment, contributing to the overall cost.

5.3.3 Reduction of graphene oxide

In this method, graphite is oxidized to form graphene oxide (GO), which is then reduced chemically or thermally to produce reduced GO. The reduction process is not perfect, and reduced GO often contains residual oxygen groups and defects, which can affect its conductivity and mechanical properties. Compared to pristine graphene, reduced GO typically has lower electrical conductivity due to the imperfections introduced during reduction. However, this is a practical and scalable method for producing graphene-like materials, especially for applications where perfect crystallinity and high conductivity are not critical, such as in composites, coatings, and some electronic devices.

5.3.4 Liquid-phase exfoliation

It is a scalable and straightforward method to produce graphene by separating individual graphene layers from bulk graphite in a liquid medium. This process involves dispersing graphite in solvents and applying external forces to break the weak van der Waals interactions that hold the graphene layers together. But compared to methods like CVD, the graphene produced via liquid-phase exfoliation may have more defects and may not achieve a perfect single-layer structure. Finding the right solvent and keeping graphene stable in dispersion can be challenging, and surfactants or additives may be required, which can affect the graphene's properties.

5.4 PHYSICAL AND CHEMICAL PROPERTIES OF GRAPHENE

Graphene's unique structure gives rise to a variety of remarkable physical and chemical properties.

5.4.1 Electrical properties

Graphene is often described as a zero-bandgap semiconductor. Its electrons behave like relativistic particles, moving at speeds much higher than electrons in traditional semiconductors like silicon. Graphene's electron mobility can reach values as high as 200,000 $cm^2/V \cdot s$, allowing for extremely fast electronic devices. Quantum Hall Effect occurs at room temperature in graphene, a phenomenon that can be utilized in quantum electronics. Functionalization or doping can adjust graphene's conductivity, making it suitable for various electronic applications.

5.4.2 Mechanical properties

Graphene is incredibly strong yet flexible. Its strength and lightness make it an ideal material for advanced composites. Graphene has a tensile strength of approximately 130 GPa, making it over 100 times stronger than steel by weight. Despite its strength, graphene can be stretched by up to 20% of its original size without breaking. Due to its flexibility and conductivity, graphene is an excellent material for wearable electronics.

5.4.3 Thermal conductivity

Graphene has superior thermal conductivity compared to other materials. With a value of around 5000 $W/m \cdot K$, graphene's ability to conduct heat surpasses even metals like copper. Heat dissipation property makes graphene valuable in electronic applications where efficient heat management is crucial, such as in processors or light-emitting diodes (LEDs).

5.4.4 Surface area and adsorption

Graphene's large surface area (up to 2630 m^2/g) makes it an excellent material for applications requiring adsorption, such as in sensors and energy storage devices. Pristine graphene repels water, although functionalization can modify this behavior. High surface area allows for the incorporation of catalytic materials, making graphene a potential catalyst support in chemical reactions.

5.5 APPLICATIONS OF GRAPHENE

5.5.1 Graphene in electronics

5.5.1.1 Graphene field-effect transistors

One of graphene's most promising applications is in field-effect transistors (FETs). The high electron mobility of graphene (up to 200,000 $cm^2/V \cdot s$)

allows graphene field-effect transistors (GFETs) to operate at significantly higher speeds than traditional silicon transistors, making them ideal for high-frequency devices. However, the lack of a bandgap in graphene makes it challenging to switch the transistor off completely, which limits its use in digital logic circuits. Researchers are working on solutions like graphene nanoribbons (GNRs) or bilayer graphene, which can introduce a small bandgap, improving their on/off ratio.

5.5.1.2 Flexible and transparent electronics

Due to its atomic thinness and exceptional flexibility, graphene is well-suited for use in flexible electronics. This includes foldable displays, wearable electronics, and electronic skins that can stretch and conform to various surfaces without losing functionality. Graphene's transparency (it absorbs only 2.3% of light) also makes it an ideal material for transparent conductive films in touchscreens, solar cells, and organic light emitting diode (OLED) displays. Graphene could replace materials like indium tin oxide (ITO), which is widely used in transparent conductors but is brittle and costly. Graphene-based films offer a more durable and flexible alternative with comparable or better performance.

5.5.1.3 Sensors and actuators

Graphene's large surface area and excellent electrical properties make it ideal for chemical and biosensors. It can detect very small changes in its environment, making it highly sensitive to gases, liquids, and biomolecules. This sensitivity has applications in medical diagnostics, environmental monitoring, and industrial sensing. Graphene-based sensors can detect molecules like glucose, proteins, or DNA, opening up possibilities for point-of-care diagnostics in healthcare. These sensors are faster and more accurate than traditional materials.

5.5.1.4 Energy storage and conversion

Graphene's high surface area and excellent conductivity also make it a strong candidate for supercapacitors, which are energy storage devices that can charge and discharge much faster than conventional batteries. Graphene supercapacitors have the potential to revolutionize energy storage by providing high power density and long cycle life. In lithium-ion batteries, graphene is used to improve the conductivity and capacity of the electrodes, leading to faster charging times and increased energy storage capacity.

5.5.1.5 Photodetectors and optoelectronics

Graphene's ability to conduct electricity and absorb light across a wide spectrum, from ultraviolet to infrared, makes it an excellent material for photodetectors and other optoelectronic devices. Graphene photodetectors can be used in high-speed communications, imaging systems, and advanced sensors. Its transparency and electrical conductivity make graphene a promising material for transparent electrodes in solar cells, improving their efficiency and reducing their production cost.

5.5.2 Graphene in nanoelectronics

Nanoelectronics involves developing electronic components at the nanometer scale (typically less than 100 nm), and graphene's atomic thickness and exceptional properties make it a powerful material in this field.

5.5.2.1 Graphene nanoribbons

They are narrow strips of graphene that, due to quantum confinement effects, can exhibit a bandgap. This allows GNRs to behave like semiconductors, which makes them useful for digital logic circuits. GNR-based transistors can achieve higher on/off ratios, addressing one of the major limitations of graphene for use in traditional electronics. GNR-based transistors have been shown to perform well at very small sizes, potentially allowing for the continued scaling down of transistor dimensions in line with Moore's Law, where silicon-based transistors face limitations.

5.5.2.2 Quantum dots and quantum computing

Graphene quantum dots (GQDs): When graphene is cut into small, nano-sized fragments, it forms graphene quantum dots. These GQDs exhibit quantum mechanical effects and can be used in applications like quantum computing, optoelectronics, and bioimaging. Quantum computing: Graphene's unique properties at the nanoscale, including the Quantum Hall Effect and its tunable electronic behavior, make it a candidate for quantum computing applications. Devices like graphene-based quantum bits (qubits) are being explored for their potential to process information much faster and more efficiently than classical computers.

5.5.2.3 Neuromorphic computing

Graphene is also being studied for use in *neuromorphic computing*—a type of computing that mimics the brain's neural network. Graphene-based transistors can function as artificial synapses, making them suitable for AI and machine learning applications that require more efficient and adaptive computing architectures.

5.5.2.4 *Nanoelectromechanical systems*

Graphene's mechanical strength and flexibility make it ideal for *nanoelectromechanical systems* (NEMS). These systems, like *resonators* and *actuators*, can operate at the nanoscale with high sensitivity and low power consumption. Graphene NEMS can be used in highly sensitive accelerometers, pressure sensors, and even nanoscale motors for medical or industrial applications.

5.5.2.5 *Tunneling devices*

In graphene-based tunnel field-effect transistors, quantum tunneling is used to switch the transistor between on and off states. This reduces power consumption compared to conventional FETs, making graphene-based TFETs attractive for ultra-low-power electronics, which are crucial for portable and battery-powered devices.

5.6 CHALLENGES IN LARGE-SCALE PRODUCTION

Despite the extraordinary properties of graphene, its large-scale production and integration into commercial devices face several challenges. These issues arise from the need to maintain high quality while ensuring affordability, scalability, and compatibility with existing technologies. The challenge in large-scale production lies in maintaining the superior properties of graphene while producing it in bulk quantities. Graphene's exceptional electrical, mechanical, and thermal properties depend heavily on its atomic structure, including layer thickness, defect density, and purity. Current methods for producing graphene, such as CVD, mechanical exfoliation, and liquid-phase exfoliation, often struggle to balance quality and scalability.

Integrating graphene into commercial devices like transistors, sensors, and energy storage systems requires seamless compatibility with existing manufacturing infrastructure. Many current production lines rely on silicon-based technologies, and graphene's integration introduces complexities due to differences in material properties, processing techniques, and assembly methods. Ensuring uniformity and connectivity between graphene and other materials, such as metals and semiconductors, is also a challenge.

Despite the progress in production techniques, the cost of producing high-quality graphene remains a significant barrier to its commercial adoption. Graphene produced via CVD is expensive due to the need for high-purity gases, specialized substrates, and precise control over growth conditions. Additionally, the equipment for mass production, such as reactors and etching tools, is costly, limiting graphene's widespread use in consumer products.

Some production methods, especially those involving chemical reduction of graphene oxide, rely on toxic solvents, hazardous chemicals, and result in chemical waste. Graphene Oxide production requires strong oxidizers like sulfuric acid and potassium permanganate, which create by-products harmful to both humans and the environment. With inhalation risks and potential cytotoxicity associated with nanoparticles, occupational exposure during production raises health concerns, especially as production scales up.

The inconsistent quality of graphene produced across different methods creates difficulties in standardizing its use in commercial devices. Without standard metrics for graphene quality (e.g., purity, layer thickness, and defect concentration), manufacturers may face issues with performance reproducibility, regulatory approval, and customer satisfaction.

Graphene's potential in transistors, sensors, energy storage systems, and flexible electronics is vast, but challenges related to production, integration, cost, and standardization remain barriers to widespread adoption. However, significant strides are being made in overcoming these obstacles through advancements in production techniques, hybrid material systems, and cost-reduction strategies. Researchers are working on eco-friendly oxidation methods, such as using hydrogen peroxide instead of harsh oxidizers like sulfuric acid. This reduces toxic by-products and improves safety. Production facilities can integrate systems for the safe handling and disposal of by-products, ensuring minimal environmental impact. Closed-loop water recycling systems can further reduce chemical runoff from the production process.

5.7 FUTURE PROSPECTS AND CONCLUSION

Future research will likely focus on improving production scalability, enhancing compatibility with existing technologies, and developing standardized quality controls to ensure that graphene can meet the demands of commercial applications. Using graphene composites or hybrid structures can help circumvent compatibility issues. For example, combining graphene with silicon-based structures or other 2D materials (e.g., transition metal dichalcogenides) may make integration more feasible without sacrificing performance. Techniques such as quantum confinement, chemical doping, and application of external electric fields can be used to induce a bandgap in graphene. While challenging, progress in this area could make graphene compatible with digital electronics without losing its advantageous properties. Collaborative efforts between research institutions, industry, and government will play a critical role in accelerating graphene's transition from the lab to the marketplace. If these challenges are addressed, graphene could usher in a new era of high-performance, flexible, and sustainable electronics, energy systems, and beyond. A future where graphene production is efficient, environmentally friendly, and economically viable will unlock its full potential, leading to revolutionary advancements in electronics, energy, and beyond.

BIBLIOGRAPHY

1. Zhang HN, Huining Zhang, Xingmao Liu, Baixiang Wang, Zhong-Bin Shi, Zhiqiang Wei, Zhi-Peng Wu,Ying Zhu, Qi Guo, Xiaolong Wang. 2023. Additive manufacturing of three-dimensional graphene-based architectures and its application in environmental treatment: a review. *Chemical Engineering Journal*. 465:142943.
2. Mohan VB, Liu D, Jayaraman K, Stam M, Bhattacharyya D. 2016. Improvements in electronic structure and properties of graphene derivatives. *Advanced Materials Letters*. 7(6):421–429.
3. Choi W, Lahiri I, Seelaboyina R, Kang YS. 2010. Synthesis of graphene and its applications: a review. *Critical Reviews in Solid State and Materials Sciences*. 35(1):52–71.
4. Lee C, Wei X, Kysar JW, Hone J. 2008. Measurement of the elastic properties and intrinsic strength of monolayer graphene. *Science*. 321(5887):385–388.
5. Min K, Aluru NR. 2011. Mechanical properties of graphene under shear deformation. *Applied Physic Letters*. 98(1):013113.
6. Suk JW, Piner RD, An J, Ruoff RS. 2010. Mechanical properties of monolayer graphene oxide. *ACS Nano*. 4(11):6557–6564.
7. Geim K, Novoselov KS. 2007. The rise of graphene. *Nature Materials*. 6(3):183–191.
8. Wu YH, Yu T, Shen ZX. 2010. Two-dimensional carbon nanostructures: fundamental properties, synthesis, characterization, and potential applications. *Journal of Applied Physics*. 108:071301.
9. Marcano DC, Kosynkin DV, Berlin JM, Sinitskii A, Sun Z, Slesarev A, Alemany LB, Lu W, Tour JM. 2010. Improved synthesis of graphene oxide. *ACS Nano*. 4(8):4806–4814.
10. Park S, An J, Jung I, Piner RD, An SJ, Li X, Velamakanni A, Ruoff RS. 2009. Colloidal suspensions of highly reduced graphene oxide in a wide variety of organic solvents. *Nano Letters*. 9(4):1593–1597.
11. Green AA, Hersam MC. 2009. Solution phase production of graphene with controlled thickness via density differentiation. *Nano Letters*. 9(12):4031–4036.
12. Hummers WS, Offeman RE. 1958. Preparation of graphitic oxide. *Journal of the American Chemical Society*. 80(6):1339.
13. Wu J, Pisula W, Müllen K. 2007. Graphenes as potential material for electronics. *Chemical Reviews*. 107(3):718–747.
14. Kim KS, Zhao Y, Jang H, Lee SY, Kim JM, Kim KS, Ahn J-H, Kim P, Choi J-Y, Hong BH. 2009. Large-scale pattern growth of graphene films for stretchable transparent electrodes. *Nature*. 457:706–710.

15. Choucair M, Thordarson P, Stride JA. 2009. Gram-scale production of graphene based on solvothermal synthesis and sonication. *Nature Nanotechnology*. 4(1):30–33.
16. Sutter PW, Flege JI, Sutter, EA. 2008. Epitaxial graphene on ruthenium. *Nature Materials*. 7:406–411.
17. Santosh KT, Sumata S, Andrzej H, Wang NN. 2020. Graphene research and their outputs: status and prospect. *Journal of Science: Advanced Materials and Devices*, 10–29.
18. Wen Y, Liu HM, Jiang XY. 2023. Preparation of graphene by exfoliation and its application in lithium-ion batteries. *Journal of Alloys and Compounds*. 961:170885.
19. Syam Sundar L, Amin Mir M, Waqar Ashraf M, Faramarz D. 2023. Synthesis and characterization of graphene and its composites for lithium-ion battery applications: a comprehensive review. *Alexandria Engineering Journal*, 78:224–245.
20. Geim AK, Novoselov KS. 2007. The rise of graphene. *Nature Materials*. 6:183–191.
21. Xiao Y, Yang JS, Wang MM, et al. 2020. Interconnected graphene scaffolds for functional gas sensors with tunable sensitivity. *Journal of Materials Science & Technology*. 58:16–23.
22. Rebecca G, Haftom W, Li HY, et al. 2023. Graphene-based nanostructures from green processes and their applications in biomedical sensors. *Advanced Industrial and Engineering Polymer Research*.7:37–53.
23. Geim AK. 2009. Graphene: status and prospects. *Science*. 324:1530–1534.
24. Scida A, Haque S, et al. 2023. Application of graphene-based flexible antennas in consumer electronic devices. *Materials Today*, 21:223–230.
25. Vishnuprasad S, Haribabu K. 2020. Synthesis of graphene encased alumina and its application as nanofluid for cooling of heat-generating electronic devices. *Powder Technology*. 363:665–675.
26. Gusynin V, Sharapov S, Carbotte J. 2007. Anomalous absorption line in the magneto-optical response of graphene. *Physical Review Letters*. 98:157402.
27. Han TH, Lee WJ, Lee DH, Kim JE, Choi EY, Kim SO. 2010. Peptide/graphene hybrid assembly into core/shell nanowires. *Advanced Materials*. 22:2060–2064.
28. Zhu Y, Murali S, Cai W, Li X, Suk JW, Potts JR, Ruoff RS. 2010. Grapheneand graphene oxide: synthesis, properties, and applications. *Advanced Materials*. 22:3906–3924.
29. Schniepp HC, Li JL, McAllister MJ, Sai H, Herrera-Alonso M, Adamson DH, et al. 2006. Functionalized single graphene sheets derived from splitting graphite oxide. *Journal of Physical Chemistry B*. 110:8535–8539.
30. Gilje S, Han S, Wang M, Wang KL, Kaner RB. 2007. A chemical Routeto graphene for device applications. *Nano Letters*. 7:3394–3398.
31. Chen H, Müller MB, Gilmore KJ, Wallace GG, Li D. 2008. Mechanically strong, electrically conductive, and biocompatible graphene paper. *Advanced Materials*. 20:3557–3561.

32. De Arco LG, Zhang Y, Zhou C. 2011. Large scale graphene by chemical vapor deposition: synthesis. *Characterization and Applications*. 196:162–184.
33. Dikin DA, Stankovich S, Zimney EJ, Piner RD, Dommett GH , Evmenenko G, et al. 2007. Preparation and characterization of graphene oxide paper. *Nature*. 448:457–460.

Chapter 6

Silicene and its derivatives

Structure, properties, modifications, and emerging applications

Akash Ramasamy, Hariharan Rajalakshmi Mohanraj, John David Thiruvadigal, and Khurshed Ahmad Shah

6.1 INTRODUCTION TO SILICENE

The scientific discovery of graphene has drawn the scientific community's attention to two-dimensional (2D) materials over the past 20 years. Bulk graphite has a sp^3-bonded carbon atom, but in graphene, it is a sp^2-bonded carbon atom, which makes three σ-bonds one π-bond in the graphene. A Dirac cone forms at the Fermi level because of these π-electrons flowing over the graphene surface as massless fermions. Due to this, graphene has a large charge mobility of 10^5 cm^2 V^{-1} s^{-1} at the Dirac point [1,2]. The unique properties of graphene attracted the researcher to work on other similar mono elemental materials called Xenes. Numerous Xenes materials, in addition to graphene, are currently being investigated experimentally as well as theoretically, including silicene [3], germanene [4], phosphorene [5], antimonene [6], etc.

Of the different types of xenes, silicene possesses sp^2–sp^3hybridization forms six Si atoms hexagonal ring, low buckling nature, Dirac cone at the fermi level, and semi-metallic nature like graphene. Due to the buckling of silicene, the two sublattices of the hexagonal ring were displaced vertically ~0.44Å [7,8]. Figure 6.1 displays the silicene's top and side views as well as the simulated band structure. In the year 2007, Guzman Verri and Yan Voon coined the term "silicene" [9].The lattice constant of silicene has been estimated to be around 3.82 Å [10].The inherent carrier mobility of silicene can attain up to $2.5\times 10^{-5}cm^2$ V^{-1} s^{-1} [11]. Silicene has good mechanical strength and a good Poisson's ratio compared to graphene [12]. Also, silicene has excellent biocompatibility and biodegradability [13].The Spin-Orbit Coupling (SOC) effect takes place in silicene because of the low buckled nature, as when we applied the electric field (EF) four different topological quantum phase transitions occurred in silicene [14].The monolayer of silicene remained stable at 1500K; nevertheless, defects may diminish its thermal stability [15].The thermoelectric figure of merit (ZT) for a silicene monolayer was determined to be 0.36 [16]. In addition, silicene possesses a substantial surface area and exceptional physicochemical features that

DOI: 10.1201/9781003512899-6

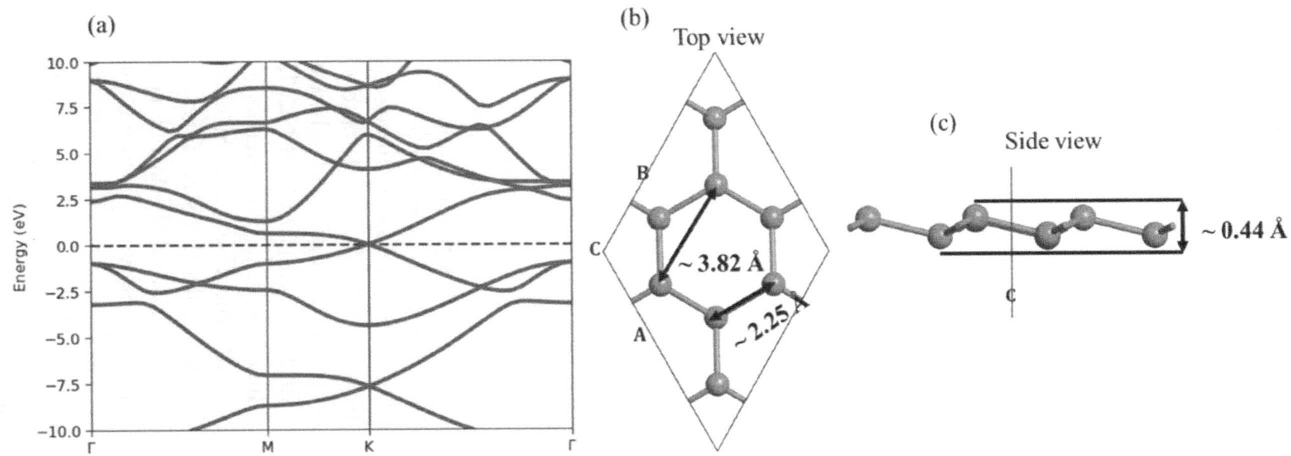

Figure 6.1 (a) Band structure of silicene, (b) top view of silicene, and (C) side view of silicene.

facilitate diverse uses [17]. Despite possessing many desired features, silicene has not been extensively utilized in electronic device applications. Different types of approaches have been employed for improving the various attributes of silicene.

6.2 EXPERIMENTAL REALIZATION OF SILICENE

Initially, silicene was epitaxially grown on a closely packed Ag (111) surface by Boubekeur Lalmi et al. in the year 2010. It was the first indication of silicene with a small buckling length of 0.22 nm, confirming the mixed sp^2–sp^3 hybridization [18]. In 2012, Patrick Vogt et al. also formed a 4× 4 silicene monolayer on the Ag (111) surface. From their study, they achieved a Si–Si bond distance of 0.22 nm which is closer to the theoretical predictions, and from the bond angle values they also confirmed the mixed sp^2–sp^3 hybridization [19].Chun-Liang Lin et al. stated the epitaxial growth of silicene on the Ag surface affects the electronic property. During epitaxial growth the Si atom made strong interaction with the silver atoms resulting in the symmetry breaking, non-Dirac fermions, and the π and π^*bands to delocalized which is confirmed by the scanning tunneling spectroscopy (STS) and density functional theory (DFT) approach [20].

Apart from the Ag surface, silicene is also grown on the ZBr_2 surface, also resulting in a similar Si–Si bond length (~2.2 nm) of theoretical values. However, the band structure reveals a band gap in silicene, probably due to large buckling [21]. Tobias G. Gill et al. also epitaxially grew the silicene on $ZrBr_2$ thin films, resulting in a metallic nature [22].Lei Meng et al. studied the construction of silicene on Ir (111) surface. Ir also interacts with the Si atom which is confirmed by the electron localization function (ELF) value, which is 0.38, but this interaction strength was much smaller for covalent interaction. This results in the electrostatic interaction of Si–Ir, so it does not affect the Si–Si bond [23]. Furthermore, silicene is epitaxially grown on insulating films apart from metal surfaces. Khalid Quertite et al. testified the growth of silicene on the insulating NaCl substrate. NaCl initially grew on the Ag (110) surface and formed islands of NaCl, which was confirmed by scanning tunneling microscopy (STM) images and a low-energy electron diffraction (LEED) pattern. Then, the Si atom grew on the NaCl islands, forming two-dimensional (2D) structures. The Si–Si bond length was measured at about 0.23 nm, and the lattice constant was determined to be 3.83 Å, which is consistent with the theoretical findings. Also, the interaction between the NaCl and the silicene was minimal compared to the interaction with metal atoms [24]. Li Tao et al. constructed a room temperature silicene field-effect transistor (FET). They performed a growth transfer fabrication technique. Initially, the Si layer was grown on top of the Ag (111)/mica substrate, and a non-reactive in situ capping technique was successfully performed to protect the silicene layer. Subsequently, the silicene was removed and placed onto the device substrate.

The silicene device has ambipolar charge transport characteristics. The attained carrier mobility was ~100 $cm^2\ V^{-1}\ s^{-1}$. Even though carrier mobility was much smaller compared to the theoretical investigation, this encapsulated delamination approach effectively produced the monolayer free-standing silicene [25].

As an alternative to the epitaxial growth of silicene, a wet chemical approach is also utilized for the synthesis of silicene. Jingjing Liu et al. first reported the few layers of silicene nanosheet (SiNS) exfoliation of $CaSi_2$. The ionic structure $CaSi_2$ included alternating layers of Si and Ca ions. The electrostatic interaction between the Ca-ions and Si-ions is stronger which makes the exfoliation process challenging. After a careful exfoliation, 0.6 nm thick SiNSs were attained. Further, the SiNSs were explored as an anode material for Li-ion batteries and the capacity values match well with the theoretical findings of bilayer silicone [26]. The SiNSs were successfully synthesized via a mild oxidation and delamination procedure based on a wet chemical methodology by Han Lin et al. The lateral size of the synthesized silicene was found to be 50 to 350 nm via transmission electron microscopy (TEM) analysis. Additionally, they examined silicene's degrading performance and discovered that it might be a viable option for biomedical applications [27]. As a tumor theranostics tool, Huican Duan et al. presented a composite incorporating a SiNS. The 2D silicene was exfoliated from the $CaSi_2$by liquid exfoliation and sonication treatment under the Ar atmosphere. Further, the MnO_x@silicene composite was done by the redox reaction among the active functional groups on the surface of the silicene and $KMnO_4$. Apart from the theragnostic studies, silicene is explored in preliminary toxicity studies which is favoring the future aspects of the study [28]. Fangfang Wang et al. explored SiNSs for drug delivery applications for tumor nanomedicine. In their study, they also followed a similar wet chemical approach to separate a free-standing silicene from the $CaSi_2$ crystals. After careful exfoliation, the silicene has a lateral size of about ~200 nm from the TEM analysis. The thickness of the SiNSs was found to be 1.5 to 3.0 nm. Their work highlights the high specific area of the silicene and the bioavailability of the material [29] (Figure 6.2).

6.3 VARIOUS APPROACHES TO TUNE PROPERTIES OF SILICENE

6.3.1 Chemical functionalization

The initial chemical functionalization is performed by hydrogenation which is adding hydrogen to each silicon atom in the structure, called silicane. In the tight binding (TB) model, the hydrogenation process opens a band gap

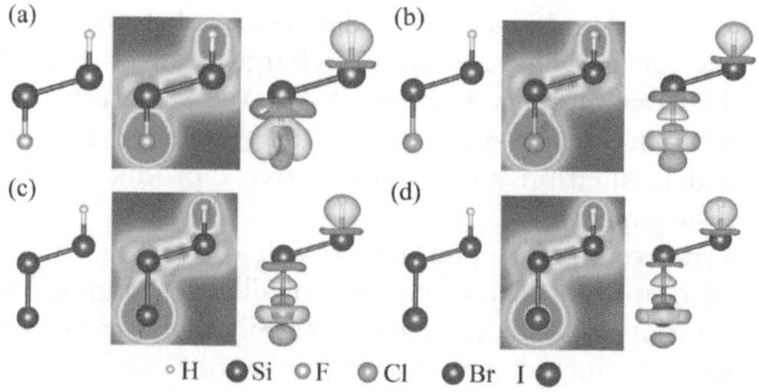

Figure 6.2 (a–d) Optimized structures, electron localization function (ELF), and charge density difference (CDD) of half-hydrogenated and half-halogenated silicene monolayer [38].

of around 2.2 eV without affecting the lattice symmetry. The electrons from the hydrogen atom occupy the vacant lowest conduction bands (CBs) of the silicene, which causes the band gap to arise [30]. Yuanyuan Pan et al. constructed silicane-based metal-oxide semiconductor field-effect transistors (MOSFETs) that have excellent gate electrostatics, transport properties, and compatibility. Also, it meets the requirements of the International Technology Roadmap for Semiconductors (ITRS) [31]. Apart from hydrogenation, halogenation can also be done to improve the characteristics of the SiNS. The halogenated silicene has a direct band gap varying from 0.54 to 2.08 eV at the semi-local Perdew-Burke-Ernzerhof (PBE) level. Also, the effective mass was found to be small which will help in electronic device applications [32]. The recent trend in chemical functionalization was asymmetric functionalization in silicene which is called Janus silicene. The direct band gap of half-hydrogenated and half-halogenated silicene, with values ranging from 1.91 to 2.66 eV, is the subject of research by Minglei Sun et al. This material is advantageous for optoelectronic applications as strain may lower the band gap value to as low as 1.11 eV [33].Aida Bao et al. report that halogenation process reduces the covalent nature between the Si–Si atom. In addition, ionicity is demonstrated by the halogen–Si bonding. The ionicity also diminished as the atomic number of halogens increased [34]. Gang Guo et al. reported a half-hydrogenated and half-alkali atom (Li and Na) functionalized SiNS. From their study, half Li atom functionalized SiNS shows a metallic nature with magnetic properties, but Na functionalized SiNS has a semiconducting nonmagnetic nature. Optical absorbance is also considerably improved in the visible spectrum [35].Pablo A. Denis studied the functionalization of H, CH_3, OH, and F into silicene symmetrically and asymmetrically, which has variation in bandgap 1.6 to 3.2 eV at the HSE level. However, in the cases of bilayer silicene, the bandgap was significantly reduced and even had a metallic nature. Fluorination affects the electronic structure of the silicene compared to other chemical functionalizations [36]. Bing Huang et al. extensively studied the chemical functionalization of O, H, F, and P with SiNS in various concentrations. In the end, they formed a new material Si_8P_4, which is an intrinsic semiconductor with a bandgap of 1.53 eV and withstanding a temperature up to 900K which opens a pathway for researchers [37].

6.3.2 Effect of doping and decoration

The electrical, optical, and magnetic attributes of silicene are greatly improved by doping or decorating different atoms in the SiNS. The single atom doping of B, N, Al, and P in the SiNS can be considered electron or hole doping. The band gap can be somewhat widened to 30 to 40 meV with the co-doping of B/N or Al/P. The careful doping of carbon in silicene can form new Dirac material like graphene which is Si_3C. In Si_3C, the Si atom contributed to the π state and the C atom contributed to the π^* state [39]. The adsorption of B, N, Al, and P can lead to different sets of results. B, N, and Al can turn the material into a nonmagnetic metal and P atom can transform the SiNS into a metal with a magnetic nature. Additionally, B, N, and P function as charge-accepting atoms, while Al acts as a charge-donating atom. All the systems with stand temperatures up to 500K, which proves the thermal stability of the system [40]. Yusheng Wang et al. studied alkali, transition, and noble metal atoms decorated SiNS for hydrogen storage applications. The binding energy of the metal atoms on the SiNS was huge, which can be helpful for avoiding the clustering on the surface. Transition metals, especially, have large binding energies. High storage capacity was found in Na, K, Mg, and Ca atom-decorated SiNS [41]. V. Ongun Özcelik and S. Ciraci reported the adsorption of Si, C, H, O, and Ti atoms on SiNS. Adding Ti at various percentages extensively tunes the SiNS electronic and magnetic behavior. It turns SiNS into a half-metal. The Si adatom forms a dumbbell-shaped structure which helps to stabilize the structure [42]. Investigating the effects of rare earth elements while doping SiNS was a comprehensive effort by Wen-Zhong Li et al. The significant magnetic moment achieved through rare earth doping has been attributed to the contribution of f-orbitals. Also, Pm and Sm doping does not affect the bonding strength of the Si–Si atom which is crucial [43].

6.3.3 Adsorption of molecules

Small organic molecule adsorption on the surface of the SiNS also can open up a band gap. Thaneshwor P. Kaloni et al. studied the change in the electronic properties of SiNS on small organic molecule adsorption. In their study, they utilized eight molecules (C_3H_6O, CH_3CN, NH_3, C_6H_6, CH_4, CH_3OH, CH_3CH_2OH, and C_7H_8). The adsorption energies range from -0.11 to -0.95 eV. The acetone adsorption opens a band gap up to 0.35 eV compared to other molecules [44]. Gas molecules such as NO_2, NH_3, and NO also can open a bandgap in SiNS [45]. The O_2molecule strongly binds to SiNS, with a value of about -6.16 eV. The O_2 dissociates into two O atoms and bonds with the Si atom. This oxidized silicene was a semiconductor [42].

6.3.4 Applying an electric field

Upon the introduction of an EF to SiNS, the material could experience the switch from a topological insulator to a band insulator. Furthermore, the application of an uneven EF on silicene enables it to adopt many states, including topological insulator, band insulator, and metal [46].Applying an out-of-plane EF significantly enhances the thermal conductivity of the SiNS without altering the structure even with low EF value. The reason for this behavior is phonon un harmonicity which is reported by Guangzhao Qin et al. [47]. The influence of a vertical EF on SiNS and its quantum transport characteristics were investigated by Zeyuan Ni et al. through the construction of the dual-gated FET. The EF expanded the band gap in SiNS, and the band gap, effective electron mass, and effective hole mass in SiNS are all linearly related to the amount of the EF. In the dual-gated FET, the EF-induced transport gap is also measured [48].

6.3.5 Applying a magnetic field

Using the TB model, Mona Abdi et al. explored the electrical and thermodynamic characteristics of planar and buckled SiNS as a function of magnetic field (MF) and doping. Applying an MF of 0 to 0.5 $g\mu_B$ B increases the heat capacity; after that, the heat capacity reduces with an increment in an external MF. The magnetic susceptibility also changes with an applied MF which makes this material suitable for spintronics applications [49]. The valley-locked spin-dependent Seebeck effect (VL-SSE) motivated by a thermal gradient was discovered by Xuechao Zhai et al. in their proposed magnetic system of silicene intercalated between two ferromagnetic (FM) dielectric layers. This effect has potential applications in logic circuits and energy devices, as it is induced by an asymmetric MF [50]. Raad Chegel and Somayeh Behzad studied the effects of applying an MF to the thermoelectric behavior of SiNS using the TB approach. Applying an MF turns silicene into a semiconductor with a direct band gap. Increasing the bias voltage further moves away the valence band maximum (VBM) and conduction band minimum (CBM) [51].

6.3.6 Effect of defects

Different types of defects are introduced in silicene, and researchers extensively studied their effect on electronic properties and stability. Shuang Li et al. reported three major forms of defects in SiNS. Single and double adatoms of silicene significantly enhance the structural stability, modifying the magnetic moment and opening a band gap [52]. Q. G. Jiang et al. studied the hydrogenation on the single vacancy- and di-vacancy-induced SiNS. Upon di-vacancy-induced SiNS, a band gap was opened, which is 0.16 eV. The dissociation of the H_2 molecule on the surface of SiNS helps to improve the I_{on}/I_{Off} ratio, which will lead to a pathway for electronic device application [53]. Junfeng Gao et al. proposed that stone Waals defects can be removed easily by an annealing process due to a small reverse transition barrier compared to graphene. The stone Waals defects also create a slight 33 meV band gap in SiNS [54].

6.3.7 Formation of heterostructure

Silicene/ZnI_2 heterostructures were formed and the structural and electronic properties were investigated by Md. Sakib Hassan et al. The heterostructure opens up a band gap of about 201 meV. Through the application of a biaxial strain and modification of the layer distance, it is possible to tune the band gap within a range of approximately zero to 318.8 meV, depending on the stacking of the heterostructure [55]. L. Shi et al. thoroughly examined the electronic, structural, and lithium storage abilities of silicene/graphene heterostructure. Forming the heterostructure benefits the silicene to withstand air atmosphere. Lithium storage properties include storage capacitance and the preservation of silicene's low migration energy. The mechanical stability is also better than the pristine silicene [56]. Ze-Bin Wu et al. sandwiched the silicene layer in between the hexagonal boron nitride (hBN) layer. The hBN/Si/hBN trilayer does not affect the electronic attributes of silicene. The sandwiching of hBN/Si/hBN/hBN/hBN/Si/hBN opens up a band gap based on the particular stacking position due to the interlayer coupling. While sandwiching the silicene in between HfO_2 clearly opens up a band gap of about 131 meV [57].

6.4 APPLICATIONS OF SILICENE

6.4.1 Nanoelectronics

Silicene's tunable bandgap and high carrier mobility make it appropriate for transistors, logic devices, and sensors. Its compatibility with silicon-based technology facilitates its integration into existing electronic systems [25,58].

6.4.2 Energy storage

Silicene's high surface area and reactivity make it a promising candidate for battery electrodes and supercapacitors. Its potential for lithium-ion and sodium-ion storage is under active investigation [26,56,59].

6.4.3 Catalysis

The chemical reactivity of silicene enhances its utility in catalysis, particularly in hydrogen evolution and CO_2 reduction reactions. Precious metal atom-doped silicene can serve as an efficient catalyst for various chemical processes [60,61].

6.4.4 Biomedical applications

Functionalized silicene has potential applications in drug delivery and biosensing due to its biocompatibility and reactivity. Its ability to interact with biological molecules makes it a promising material for medical diagnostics and therapeutics [27–29].

6.4.5 Optoelectronics

Silicene's optical properties render it a contender for photodetectors, solar cells, and light-emitting devices. Its tunable bandgap allows for wavelength-specific applications, broadening its optoelectronic utility [62–64] (Figure 6.3).

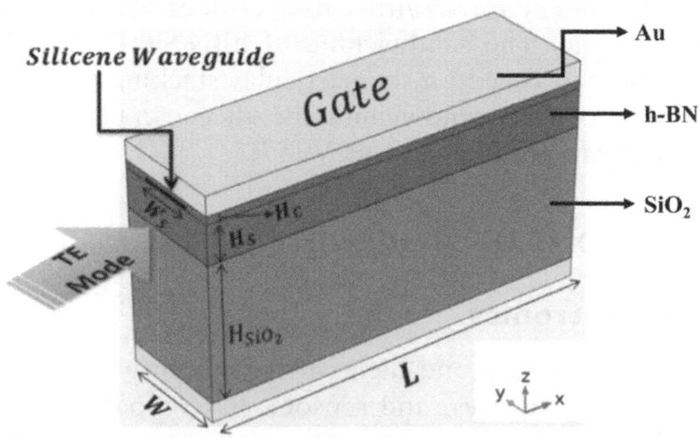

Figure 6.3 The 3D diagram of optical MOSFET [64].

6.5 DERIVATIVES OF SILICENE

6.5.1 Silicene nanotubes

Similar to carbon nanotubes (CNTs), silicenenanotubes (SiNTs) can be formed by rolling a silicon layer. The SiNT acts as a topological insulator in this setup, where the application of an electric field (EF) causes the metallic region to delineate the topological insulator from the band insulator. This characteristic can make its way into the field of spintronics [65]. Nonequilibrium molecular dynamics (NEMD)-based calculations reveal that SiNT exhibits length-dependent thermal conductance. Also, SiNT shows good mechanical stability under strain and grain boundaries. The grain boundaries significantly reduce the thermal conductivity due to Kapitza's thermal resistance. Additionally, the thermal conductance was increased when applying a tensile strain [66]. Raad Chegel studied the nonlinear optical properties of CNTs and SiNT. Varying the radius of the nanotubes affects the optical properties. Due to the low energy peaks and higher intensities, SiNT has an edge in practical applications over CNT [67]. To investigate how factors such as channel length, chirality, and tube diameter affect the device's properties, Faranak Salimian and Daryoosh Dideban designed a silicene nanotube field-effect transistor (SiNTFET). An increasing applied EF caused the band gap of SiNT to widen. Additionally, the channel length significantly affects the I_{on}/I_{Off} ratio, and increment in the diameter of the band gap reduces. Therefore, it affects the off current. Compared to carbon nanotube field-effect transistor (CNTFET), SiNFET performs well [68].

6.5.2 Silicene nanoribbon

The optimal technique for modifying the characteristics of silicene was segmenting it into a nanoribbon. Silicene has a strong anisotropic nature which makes this into an orientation-dependent characteristic. The quantum confinement in the particular direction of silicene makes nanoribbons interesting. Based on the cutting angle, the two major divisions in the nanoribbon are zigzag silicene nanoribbon (ASiNR) and armchair silicene nanoribbon (ASiNR), which are shown in Figure 6.4.

6.5.2.1 Zigzag silicene nanoribbon (ZSINR)

The size of the nanoribbons is notified by the number of Si atoms in the edge. The edges on the transport direction have dangling bonds, which causes significant disorders in the structure and an anti-ferromagnetic (AFM) and metallic nature. Upon hydrogen termination, the condition was changed [69]. Also, the pristine ZSiNR has poor stability, which can be enhanced

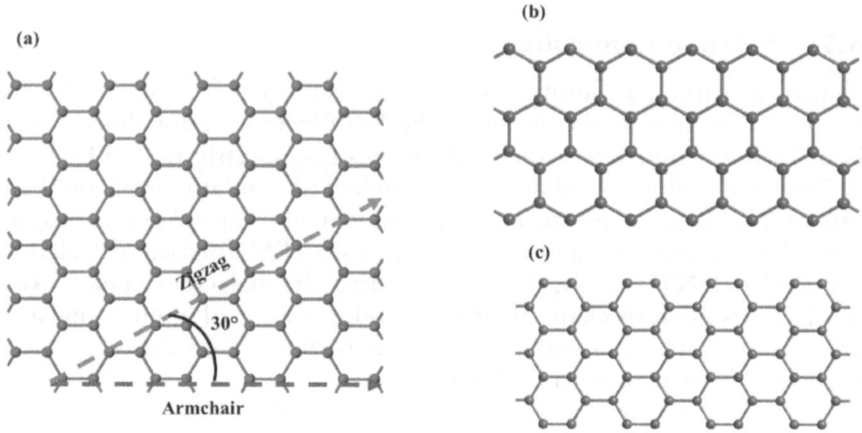

Figure 6.4 (a) Illustration of the zigzag and armchair SiNR cutting direction and angle, (b) zigzag SiNR, and (C) armchair SiNR.

by hydrogenation or other functionalization at the dangling bonds. Hydrogenation on both edges results in a tiny indirect bandgap of 0.036 eV. Upon double functionalization at the edges, it can open a direct bandgap of ~0.02 to 0.175 eV in ZSiNR. While doping the B or N in ZSiNR, the band gap remains zero [70]. Yu-Ling Song et al. reported the Si–Si bond length at the edge shorter than the inner ones. Also, the Si–H bond length was 1.50 Å in both ZSiNR and ASiNR; in addition, Si–H bond was ionic due to the large electronegativity difference. A single hydrogenated N-ZSiNR (where N=4, 5, 6, etc.) exhibits a zero band gap [71].

6.5.2.2 Armchair silicene nanoribbon (ASINR)

The cutting angle difference between the zigzag and armchair SiNR was about 30°. The ASiNR also had structural deformation in a pristine state [69]. Here also the stability was increased by hydrogenation or any other functionalization. Also, ASiNR has unique electronic properties like armchair graphene nanoribbon (AGNR), which is width-dependent electronic band gap, unlike having a zero bandgap after a 4-atom width in ZSiNR. Based on the width of the ASiNR, it can be categorized into three sets which are 3p–1, 3p, and 3p+1where p (=1, 2, 3, etc.) is the integer. About each set, the band gap value rises. Moreover, the band gap value decayed as the width increased in each group [69,70].

6.6 APPLICATIONS OF SINR

The width-dependent and other characteristics of silicene nanoribbons attracted the research community to work on various applications which is shown in Figure 6.5.

6.6.1 Biosensing and gas sensing

The biosensing and harmful gas sensing applications of ASiNR were investigated in depth because of its width-dependent band gap, which causes a change in electronic characteristics upon adsorption. Akash Ramasamy et al. thoroughly investigated the sensing performance of chronic obstructive pulmonary disease (COPD) breath biomarker ($C_6H_{12}O$) and lung cancer biomarkers (C_3H_8O and C_3H_6O) on the pristine, defect induced, B, C, and N-doped 7-atom width ASiNR. From their findings, the p-block elements doping and the defects can significantly boost the sensitivity of ASiNR. Also, the adsorption energy value for $C_6H_{12}O$ was reported for N-ASiNR compared to other materials reported early [72,73]. Shazia Showket et al. reported Al-doped, P-doped, Al/P co-doped, and vacancy defected 7-ASiNR for sensing volatile organic compounds (VOCs) such as CH_2O and C_2H_4O. A two-probe device model was constructed to study the transport abilities and calculate the sensitivity of ASiNR-based structures. Al/P co-doped ASiNR has a large current value at the bias voltage of 2 V. Additionally, it shows huge sensitivity values of around 68% and77% for CH_2O and C_2H_4O, respectively [74]. Apart from VOC sensing, SiNR is also used in DNA sequencing which is the recognition of DNA methylation. Qingxin Li et al. studied the interaction of eight nucleobases toward the ZSiNR. Apart from adenine, all other nucleobases chemically interacted with the ZSiNR. These nucleobases can be identified with help of "high-low" strategy from the I to V characteristics [75]. Sara Alesheikh et al. investigated a ZSiNR as a biosensor for DNA sequencing using DFT with nonequilibrium Green's function (NEGF) approach. The ZSiNR is used as a sensing medium for detecting adenine, guanine, thymine, and cytosine. Based on their findings, guanine has a strong interaction with higher sensitivity [76].

Toxic gas detection is one of the major problems in the current scenario due to global warming and the greenhouse effect. The silicene nanoribbon detects various toxic gases. S. M. Aghaei et al. investigated the sensing attributes of the ASiNR for a wide variety of harmful gases, such as N_2, NO, NO_2, NH_3, CO, CO_2, CH_4, SO_2, and H_2S. The pristine ASiNR has good sensitivity toward CO, NH_3, and H_2S gas molecules. However, pristine ASiNR is unsuitable for detecting the NO and NO_2 due to the strong interaction. Also, the impact of B and N atoms doped ASiNR and dangling bond defected ASiNR was studied to detect toxic gases. From that, B-doped ASiNR can be a good sensor for N_2 detection [77]. The effect of a

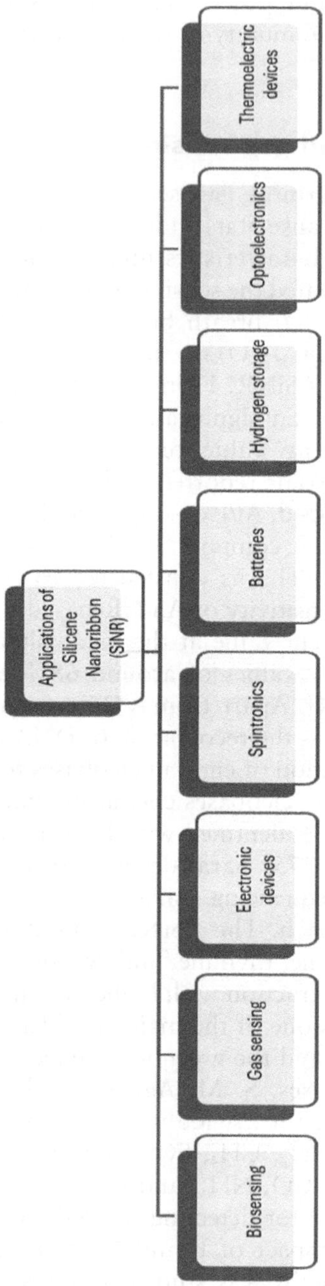

Figure 6.5 The use of silicene nanoribbon (SiNR) across the wide area of research.

single vacancy in ASiNR for toxic gas detection was investigated by Gurleen Kaur Walia et al. The toxic gases split into three groups based on core components such as carbon (C)-based (CO, CO_2, and CH_4), sulfur (S)-based (H_2S and SO_2), and nitrogen (N)-based (NO_2 and NO) gas molecules. CO_2 and CH_4gas molecules weakly adsorbed on the vacancy-induced ASiNR. However, CO has made strong chemical bonding with Si atoms resulting in huge sensitivity and very large recovery time. In these cases, S-based toxic gas adsorption on single-vacancy ASiNR had strong chemical interaction. Even the H_2S molecule dissociates into two parts. For N-based gases in both pristine and vacancy created ASiNR, they made a strong chemical binding [78–80].

6.6.2 Electronic device applications

Jun Kang et al. reported the transport properties of odd (N=7) and even (N=6) atom width ZSiNR using the DFT approach along with the NEGF scheme. The 7-ZSiNR does not have a mirror plane which leads to strong conducting nature. This means that the current value is directly proposed to the applied bias voltage. However, in the case of 6-ZSiNR, the σ mirror plane was present at the Fermi level, leading to the transmission gap in 6-ZSiNR. Here we obtain only a small amount of current. This behavior leads to the switching of low magnetoresistance (MR) to higher MR up to 1×10^{-7}%. This phenomenon can be applied to logic gate devices [81]. Dongqing Zou et al. studied the transport abilities of single- and double-hydrogenated ZSiNR as well as the O or OH-terminated with symmetric and asymmetric type in 6-atom width ZSiNR. The functionalization of different compounds had a strong impact on the transport characteristics. Under applied bias, the π and π^* sub bands are coupled with each other, contributing to the conductance and reason for the increase in current [82]. Fei Wan et al. examined the electronic properties and transport abilities of ZSiNR with the effect of line defect by TB approach along with NEGF. Due to the impact of a line defect, the sublattice symmetry was breaking and the flat band bent upward or downward. The line defect present in the middle improves the conductance of the ZSiNR. The key factors affecting the conductance were sublattice potential, the on-site energy, and the spin–orbit coupling (SOC) effect [83].

Hong Li et al. constructed ASiNR-based FET to study the transport characteristics using DFT with NEGF approach. The N-ASiNR (N= 6, 7, 8, and 10) is connected to the metallic ZSiNR electrodes. The ASiNR FET has a high I_{on}/I_{off} ratio of about 10^6 along with a bipolar gate effect. The output characteristics are enhanced by increasing the length of the ASiNR. Additionally, the saturation nature of ASiNR-based FET makes it better than AGNR-based FET [84]. Mu Wen Chuan et al. studied the non-ballistic phonon scattering effects in the SiNR-based FET devices by incorporating the ballistic top-of-the-barrier (ToB) nanotransistor model. The 7-ASiNR has the highest I_{on}/I_{off} of about 3.32 × 10^5 among the 3p+1 family which is crucial for the device application. The high I_{on}/I_{off} attained the largest length chosen for the study which is 10 Å. By incorporating the phonon scattering effect, the I_{on}/I_{off} significantly reduced [85]. Armin Gooran-Shoorakchaly et al. construct the electrostatically doped ASiNR-based FET to study the transport properties by using numerical simulations. Upon electrostatically doping ASiNR, FET exhibits a large I_{on}/I_{off} ratio of 2.4 × 10^4. However, extended channel electrostatically doped ASiNR FET has an even higher I_{on}/I_{off} ratio of 3.8 × 10^5 [86]. Mirko Poljak builds the ASiNR-based MOSFET with the 15-nm-gatelength. This ASiNR MOSFET competes with the other materials and benchmarked basic characteristics set by International Technology Roadmap for Semiconductors (IRDS) [87].

6.6.3 Spintronic application

Chengyong Xu et al. reported the quantum transport characteristics of N-ZSiNR (3, 4, 5, 6, 7, and 8) with semi-infinite silicene electrodes using DFT with NEGF approach. The calculated MR for the N-ZSiNR varies from 261% to 1960%, which can result in the practical application of spintronics application [88]. Mubashir A. Kharadi and Ibrahim Mahariq thoroughly investigated the spin caloritronic device characteristics by forming a heterojunction with H-terminated ZSiNR and H terminated with B/N-co doped ZSiNR. Without applying a bias voltage, a device operates, and spin transport takes place only by the thermal gradient. The MR attained a huge value of –8× $10^{-9}$and the positive negative transition takes place between 160 and180K. These switching properties can be helpful for spin-filtering applications [89]. Dang-Qi Fang et al. testified that the formation energy of single- and double-hydrogenated ZSiNR depends on the width. Also, a double-hydrogenated system possesses the AFM semiconducting nature. In addition, the doping of P and N in the hydrogenated ZSiNR makes the material half-metal [90]. Gang Guo et al. studied the effect of H, F, and B atoms at the edge of the 6-ZSiNR using the DFT approach. The 6-ZSiNR exhibits the half metallic nature, while half halogenated or fluorinated and half functionalized with B atoms. Also, half hydrogenated and half B atom adsorbed 6-ZSiNR holds a large magnetic moment, which provides a path to nanodevices [91]. X.Q. Deng et al. investigated the electronic and magnetic properties of half surface hydrogenated ZSiNR, fully surface hydrogenated ZSiNR heterostructure, and heterostructure of ZSiNR and half-surface hydrogenated ZSiNR. The ferromagnetism (FM) can be introduced in the ZSiNR by half-surface hydrogenated ZSiNR. Then, heterostructure exhibits excellent spin-filtering effect along with parallel and anti-parallel spin configuration. The negative differential region (NDR) plays a crucial role in anti-parallel spin. These research finding suggests potential for ZSiNR in future spintronic device applications [92]. Saurabh Kharwar et al. studied the structural and electronic properties of symmetrically and asymmetrically hydrogenated and fluorinated ZSiNR. The structural stability is not affected by the symmetrical and asymmetrical hydrogenation and fluorination. Also, bipolar magnetic semiconductors and half-metallic behavior were identified from the calculation of the electronic structure. Furthermore, the transport studies of the ZSiNR utilising a two-probe device model demonstrate negative differential resistance (NDR) that is attributed to the spin-down current [93]. Overall, passivation on the edges paves the path for the spintronic application using ZSiNR.

6.6.4 Hydrogen storage

One of the cleanest energy sources is hydrogen, and there has long been a hunt for materials to store hydrogen. Yuliang Mao et al. explored the electrical and magnetic characteristics, as well as the hydrogen storage capabilities, of ZSiNR. At the dangling bond edges, mono and di hydrogenation drastically alter the magnetic behavior of the ZSiNR [94].

6.6.5 Batteries

Using a first principles study, ZSiNR was examined as a Li storage material by adding Li as an adatom to its edge. The symmetrical incorporation of Li atoms at the edge of the ZSiNR changes the FM state to the AFM state. The Li atoms have an ionic interaction with the Si atoms with reasonable charge transfer [95].

6.6.6 Thermoelectric devices

Long Xu et al. studied the impact of Ti adatom on the structural, electronic, and thermoelectric properties of the ZSiNR using DFT with the NEGF approach. The single Ti adatom stays at the center of the ZSiNR rather than the edges. Upon adding two adatoms symmetrically, it does not affect the giant MR. Moreover, spin thermoelectric attributes improved by adding a Ti adatom [96].

6.6.7 Optoelectronic applications

Optoelectronic devices play a crucial role in communication devices, making it necessary to find a suitable material. Mubashir A. Kharadi et al. reported Li and Cl co-decorated ZSiNR as a photodetector device application. The calculated performance parameter of the ZSiNR has been compared with established materials and found to outperform other materials [97]. Z. Abbasi Azad et al. explored the electronic and transport abilities of Al and P-doped ASiNR. Doping impurities like Al and N change the intrinsic semiconducting nature of ASiNR into n-type and p-type semiconductors. The effective electron masses increase by the impact of doping. The I–V characteristics also became nonlinear after the threshold value. Those attributes made ASiNR a strong candidate for optoelectronic application and similar device applications [98–100].

6.7 CONCLUSION

Silicene, one of the Dirac materials with a low buckled nature, also has many interesting features such as the quantum spin Hall effect, giant MR, and biocompatibility. From the experimental point of view, silicene still needs to be explored more. The semi-metallic nature restricts silicene in various device applications. To overcome the issues, various approaches have been employed for the past two decades. The surface functionalization with the hydrogen atom and halogen atom opens up a wide band gap in silicene, which is explored in catalytic applications due to its unique optical properties. Similarly, doping, defects, small-molecule adsorption, applying an EF, applying an MF, forming a heterostructure, and making SiNS into a 1D nanotube or nanoribbon were also explored. Among various approaches to tuning the properties of silicene, cutting it into a nanoribbon made admirable attributes. The effect of quantum confinement and width-dependent characteristics of SiNR as notable applicant for electronic device applications. Nevertheless, silicene and the derivatives of the silicene need to be investigated.

REFERENCES

[1] S.M.-M. Dubois, Z. Zanolli, X. Declerck, J.-C. Charlier, Electronic properties and quantum transport in Graphene-based nanostructures, *Eur. Phys. J. B* 72 (2009) 1–24. https://doi.org/10.1140/epjb/e2009-00327-8

[2] A.H. Castro Neto, F. Guinea, N.M.R. Peres, K.S. Novoselov, A.K. Geim, The electronic properties of graphene, *Rev. Mod. Phys.* 81 (2009) 109–162. https://doi.org/10.1103/RevModPhys.81.109

[3] H. Oughaddou, H. Enriquez, M.R. Tchalala, H. Yildirim, A.J. Mayne, A. Bendounan, G. Dujardin, M. Ait Ali, A. Kara, Silicene, a promising new 2D material, *Prog. Surf. Sci.* 90 (2015) 46–83. https://doi.org/10.1016/j.progsurf.2014.12.003

[4] A. Acun, L. Zhang, P. Bampoulis, M.V. Farmanbar, A. van Houselt, A.N. Rudenko, M. Lingenfelder, et al., Germanene: The germanium analogue of graphene, *J. Phys. Condens. Matter.* 27 (2015) 443002. https://iopscience.iop.org/article/10.1088/0953-8984/27/44/443002/

[5] A. Carvalho, M. Wang, X. Zhu, A.S. Rodin, H. Su, A.H. Castro Neto, Phosphorene: From theory to applications, *Nat. Rev. Mater.* 1 (2016) 1–16. https://doi.org/10.1038/natrevmats.2016.61

[6] X. Wang, J. Song, J. Qu, Antimonene: From experimental preparation to practical application, *Angew. Chem. Int. Ed. Engl.* 58 (2019) 1574–1584. https://onlinelibrary.wiley.com/doi/abs/10.1002/anie.201808302
[7] N.D. Drummond, V. Zólyomi, V.I. Fal'ko, Electrically tunable band gap in silicene, *Phys. Rev. B.* 85 (2012) 075423. https://doi.org/10.1103/PhysRevB.85.075423
[8] M. Houssa, A. Dimoulas, A. Molle, Silicene: A review of recent experimental and theoretical investigations, *J. Phys. Condens. Matter.* 27 (2015) 253002. https://doi.org/10.1088/0953-8984/27/25/253002
[9] G.G. Guzmán-Verri, L.C. Lew Yan Voon, Electronic structure of silicon-based nanostructures, *Phys. Rev. B.* 76 (2007) 075131. https://doi.org/10.1103/PhysRevB.76.075131
[10] L. C. Lew Yan Voon, J. Zhu, U. Schwingenschlögl, Silicene: Recent theoretical advances, *Appl. Phys. Rev.* 3 (2016) 040802. https://pubs.aip.org/aip/apr/article-abstract/3/4/040802/124066/
[11] Z.-G. Shao, X.-S. Ye, L. Yang, C.-L. Wang, First-principles calculation of intrinsic carrier mobility of silicene, *J. Appl. Phys.* 114 (2013) 093712. https://doi.org/10.1063/1.4820526
[12] Q. Peng, X. Wen, S. De, Mechanical stabilities of silicene, *RSC Adv.* 3 (2013) 13772–13781. https://doi.org/10.1039/C3RA41347K
[13] Y. You, C. Yang, X. Zhang, H. Lin, J. Shi, Emerging two-dimensional silicene nanosheets for biomedical applications, *Mater. Today Nano* 16 (2021) 100132. https://doi.org/10.1016/j.mtnano.2021.100132
[14] M. Ezawa, Valley-polarized metals and quantum anomalous hall effect in silicene, *Phys. Rev. Lett.* 109 (2012) 055502. https://doi.org/10.1103/PhysRevLett.109.055502
[15] G.R. Berdiyorov, F.M. Peeters, Influence of vacancy defects on the thermal stability of silicene: A reactive molecular dynamics study, *RSC Adv.* 4 (2013) 1133–1137. https://doi.org/10.1039/C3RA43487G
[16] K. Yang, S. Cahangirov, A. Cantarero, A. Rubio, R. D'Agosta, Thermoelectric properties of atomically thin silicene and germanene nanostructures, *Phys. Rev. B* 89 (2014) 125403. https://doi.org/10.1103/PhysRevB.89.125403
[17] L. Ma, X. Song, Y. Yu, Y. Chen, Two-dimensional silicene/silicon nanosheets: An emerging silicon-composed nanostructure in biomedicine, *Adv. Mater.* 33 (2021) 2008226. https://doi.org/10.1002/adma.202008226
[18] B. Lalmi, H. Oughaddou, H. Enriquez, A. Kara, S. Vizzini, B. Ealet, B. Aufray, Epitaxial growth of a silicene sheet, *Appl. Phys. Lett.* 97 (2010) 223109. https://doi.org/10.1063/1.3524215
[19] P. Vogt, P. De Padova, C. Quaresima, J. Avila, E. Frantzeskakis, M.C. Asensio, A. Resta, B. Ealet, G. Le Lay, Silicene: Compelling experimental evidence for graphene like two-dimensional silicon, *Phys. Rev. Lett.* 108 (2012) 155501. https://doi.org/10.1103/PhysRevLett.108.155501
[20] C.-L. Lin, R. Arafune, K. Kawahara, M. Kanno, N. Tsukahara, E. Minamitani, Y. Kim, M. Kawai, N. Takagi, Substrate-induced symmetry breaking in silicene, *Phys. Rev. Lett.* 110 (2013) 076801. https://doi.org/10.1103/PhysRevLett.110.076801

[21] A. Fleurence, R. Friedlein, T. Ozaki, H. Kawai, Y. Wang, Y. Yamada-Takamura, Experimental evidence for epitaxial silicene on diboride thin films, *Phys. Rev. Lett.* 108 (2012) 245501. https://doi.org/10.1103/PhysRevLett.108.245501

[22] T.G. Gill, A. Fleurence, B. Warner, H. Prüser, R. Friedlein, J.T. Sadowski, C.F. Hirjibehedin, Y. Yamada-Takamura, Metallic atomically-thin layered silicon epitaxially grown on silicene/ZrB2, *2D Mater.* 4 (2017) 021015.https://doi.org/10.1088/2053-1583/aa5a80

[23] L. Meng, Y. Wang, L. Zhang, S. Du, R. Wu, L. Li, Y. Zhang, G. Li, H. Zhou, W.A. Hofer, H.-J.Gao, Buckled silicene formation on Ir(111), *Nano Lett.* 13 (2013) 685–690. https://doi.org/10.1021/nl304347w

[24] K. Quertite, H. Enriquez, N. Trcera, Y. Tong, A. Bendounan, A.J. Mayne, G.Dujardin, P. Lagarde, A. El kenz, A. Benyoussef, Y.J. Dappe, A. Kara , H. Oughaddou, Silicene nanoribbons on an insulating thin film, *Adv. Funct. Mater.* 31 (2021) 2007013.https://doi.org/10.1002/adfm.202007013

[25] L. Tao, E. Cinquanta, D. Chiappe, C. Grazianetti, M. Fanciulli, M. Dubey, A. Molle, D. Akinwande, Silicene field-effect transistors operating at room temperature, *Nat. Nanotechnol.* 10 (2015) 227–231. www.nature.com/articles/nnano.2014.325 (accessed January 10, 2025).

[26] J. Liu, Y. Yang, P. Lyu, P. Nachtigall, Y.Xu, Few-layer silicene nanosheets with superior lithium-storage properties, *Adv. Mater.* 30 (2018) 1800838. https://doi.org/10.1002/adma.201800838

[27] H. Lin, W. Qiu, J. Liu, L. Yu, S. Gao, H. Yao, Y. Chen, J. Shi, Silicene: Wet-chemical exfoliation synthesis and biodegradable tumor nanomedicine, *Adv. Mater.* 31 (2019) 1903013. https://doi.org/10.1002/adma.201903013

[28] H. Duan, H. Guo, R. Zhang, F. Wang, Z. Liu, M. Ge, L. Yu, H. Lin, Y. Chen, Two-dimensional silicene composite nanosheets enable exogenous/endogenous-responsive and synergistic hyperthermia-augmented catalytic tumor theranostics, *Biomaterials* 256 (2020) 120206. https://doi.org/10.1016/j.biomaterials.2020.120206

[29] F. Wang, H. Duan, R. Zhang, H. Guo, H. Lin, Y. Chen, Potentiated cytosolic drug delivery and photonic hyperthermia by 2D free-standing silicene nanosheets for tumor nanomedicine, *Nanoscale* 12 (2020) 17931–17946. https://doi.org/10.1039/D0NR05214K

[30] G.G. Guzmán-Verri, L.C. Lew Yan Voon, Band structure of hydrogenated Si nanosheets and nanotubes, *J. Phys. Condens. Matter* 23 (2011) 145502. https://doi.org/10.1088/0953-8984/23/14/145502

[31] Y. Pan, J. Dai, L. Xu, J. Yang, X. Zhang, J. Yan, J. Li, B. Shi, S. Liu, H. Hu, M. Wu, J. Lu, Sub-5-nm monolayer silicane transistor: A first-principles quantum transport simulation, *Phys. Rev. Appl.* 14 (2020) 024016. https://doi.org/10.1103/PhysRevApplied.14.024016

[32] W.-B. Zhang, Z.-B. Song, L.-M. Dou, The tunable electronic structure and mechanical properties of halogenated silicene: A first-principles study, *J. Mater. Chem.* C3 (2015) 3087–3094. https://doi.org/10.1039/C4TC02758B

[33] M. Sun, Q. Ren, S. Wang, J. Yu, W. Tang, Electronic properties of Janus silicene: New direct band gap semiconductors, *J. Phys. Appl. Phys.* 49 (2016) 445305. https://doi.org/10.1088/0022-3727/49/44/445305

[34] A. Bao, X. Li, X. Guo, H. Yao, M. Chen, Tuning the structural, electronic, mechanical and optical properties of silicene monolayer by chemical functionalization: A first-principles study, *Vacuum* 203 (2022) 111226. https://doi.org/10.1016/j.vacuum.2022.111226

[35] G. Guo, Y. Xu , G. Guo, Janus-functionalization induced magnetism and improved optoelectronic properties in two-dimension silicene and germanene: Insights from first-principles calculations, *J. Phys. Condens. Matter.* 35 (2023) 335501. https://doi.org/10.1088/1361-648X/acd50d

[36] P.A. Denis, Stacked functionalized silicene: A powerful system to adjust the electronic structure of silicene, *Phys. Chem. Chem. Phys.* 17 (2015) 5393–5402. https://doi.org/10.1039/C4CP05331A

[37] B. Huang, H.J. Xiang, S.-H. Wei, Chemical functionalization of silicene: Spontaneous structural transition and exotic electronic properties, *Phys. Rev. Lett.* 111 (2013) 145502. https://doi.org/10.1103/PhysRevLett.111.145502

[38] G. Guo, J. Liu, Y. Xu, G. Guo, S. Tan, Chemical functionalization induced photocatalytic performance for water splitting of silicene: A first-principles investigation, *Colloids Surf. A Physicochem. Eng. Asp.* 667 (2023) 131379. https://doi.org/10.1016/j.colsurfa.2023.131379

[39] N. Yin, Y. Dai, W. Wei, B. Huang, Electronic structure engineering in silicene via atom substitution and a new two-dimensional Dirac structure Si3C, *Physica E Low Dimens. Syst. Nanostruct.* 98 (2018) 39–44. https://doi.org/10.1016/j.physe.2017.12.024

[40] J. Sivek, H. Sahin, B. Partoens, F.M. Peeters, Adsorption and absorption of boron, nitrogen, aluminum, and phosphorus on silicene: Stability and electronic and phonon properties, *Phys. Rev. B* 87 (2013) 085444. https://doi.org/10.1103/PhysRevB.87.085444

[41] Y. Wang, R. Zheng, H. Gao, J. Zhang, B. Xu, Q. Sun, Y. Jia, Metal adatoms-decorated silicene as hydrogen storage media, *Int. J. Hydrogen Energy* 39 (2014) 14027–14032. https://doi.org/10.1016/j.ijhydene.2014.06.164

[42] V. Ongun Özçelik, S. Ciraci, Local reconstructions of silicene induced by adatoms, *J. Phys. Chem. C* 117 (2013) 26305–26315. https://pubs.acs.org/doi/10.1021/jp408647t (accessed January 10, 2025).

[43] W.-Z. Li, M.-Y. Liu, L. Gong, Q.-Y. Chen, C. Cao, Y. He, Emerging various electronic and magnetic properties of silicene by light rare-earth metal substituted doping, *Superlattices Microstruct.* 148 (2020) 106712. https://doi.org/10.1016/j.spmi.2020.106712

[44] T.P.Kaloni, G. Schreckenbach, M.S. Freund, Large enhancement and tunable band gap in silicene by small organic molecule adsorption, *J. Phys. Chem. C* 118 (2014) 23361–23367. https://doi.org/10.1021/jp505814v

[45] W.Hu , N. Xia, X. Wu, Z. Li, J. Yang, Silicene as a highly sensitive molecule sensor for NH3, NO and NO2, *Phys. Chem. Chem. Phys.* 16 (2014) 6957. https://doi.org/10.1039/c3cp55250k

[46] M. Ezawa, A topological insulator and helical zero mode in silicene under an inhomogeneous electric field, *New J. Phys.* 14 (2012) 033003. https://doi.org/10.1088/1367-2630/14/3/033003

[47] G. Qin, Z. Qin, S.-Y. Yue, Q.-B. Yan, M. Hu, External electric field driving the ultra-low thermal conductivity of silicene, *Nanoscale* 9 (2017) 7227–7234. https://doi.org/10.1039/C7NR01596H

[48] Z. Ni, Q. Liu, K. Tang, J. Zheng, J. Zhou, R. Qin, Z. Gao, D. Yu, J. Lu, Tunable bandgap in silicene and germanene, *Nano Lett.* 12 (2012) 113–118. https://doi.org/10.1021/nl203065e

[49] M. Abdi, E. Norian, B. Astinchap, Effect of external magnetic field and doping on electronic and thermodynamic properties of planer and buckled silicene monolayer, *Sci. Rep.*12 (2022) 22277. www.nature.com/articles/s41598-022-26353-1 (accessed January 10, 2025).

[50] X. Zhai, Y.-T. Wang, R. Wen, S.-X. Wang, Y. Tian, X. Zhou, W. Chen, Z. Yang, Valley-locked thermospin effect in silicene and germanene with asymmetric magnetic field induced by ferromagnetic proximity effect, *Phys. Rev. B* 97 (2018) 085410. https://doi.org/10.1103/PhysRevB.97.085410

[51] R. Chegel, S. Behzad, Controlling the thermoelectric behaviors of biased silicene via the magnetic field: Tight binding model, *Physica E Low Dimens. Syst. Nanostruct.* 135 (2022) 114945. https://doi.org/10.1016/j.physe.2021.114945

[52] S. Li, Y. Wu, Y. Tu, Y. Wang, T. Jiang, W. Liu, Y. Zhao, Defects in silicene: Vacancy clusters, extended line defects and di-adatoms, *Sci. Rep.* 5 (2015) 7881. https://doi.org/10.1038/srep07881

[53] Q.G. Jiang, W.C. Wu, J.F. Zhang, Z.M. Ao, Y.P. Wu, H.J. Huang, Defections induced hydrogenation of silicene: A density functional theory calculation study, *RSC Adv.* 6 (2016) 69861–69868. https://doi.org/10.1039/C6RA11885B

[54] J. Gao, J. Zhang, H. Liu, Q. Zhang, J. Zhao, Structures, mobilities, electronic and magnetic properties of point defects in silicene, *Nanoscale* 5 (2013) 9785–9792. https://doi.org/10.1039/C3NR02826G

[55] Md. S. Hassan, Md. S. Islam, J. Park, Silicene/ZnI2 van der Waals heterostructure: Tunable structural and electronic properties, *Nanotechnology* 32 (2021) 305707. https://doi.org/10.1088/1361-6528/abf9c6

[56] L. Shi, T.S. Zhao, A. Xu, J.B. Xu, Ab initio prediction of a silicene and graphene heterostructure as an anode material for Li- and Na-ion batteries, *J. Mater. Chem. A* 4 (2016) 16377–16382. https://doi.org/10.1039/C6TA06976B

[57] Z.-B. Wu, Y.-Y. Zhang, G. Li, S. Du, H.-J. Gao, Electronic properties of silicene in BN/silicene van der Waals heterostructures, *Chin. Phys. B* 27 (2018) 077302. https://doi.org/10.1088/1674-1056/27/7/077302

[58] M.P. Lima , A. Fazzio, A.J.R. daSilva, Silicene-based FET for logical technology, *IEEE Electron Device Lett.* 39 (2018) 1258–1261. https://doi.org/10.1109/LED.2018.2848640

[59] J. Zhu, U. Schwingenschlögl, Silicene for Na-ion battery applications, *2D Mater.* 3 (2016) 035012. https://iopscience.iop.org/article/10.1088/2053-1583/3/3/035012/meta (accessed January 11, 2025).

[60] S. Zhou, W. Pei, J. Zhao, A. Du, Silicene catalysts for CO2 hydrogenation: The number of layers controls selectivity, *Nanoscale* 11 (2019) 7734–7743. https://doi.org/10.1039/C9NR01336A

[61] W.-Z. Li, M.-Y. Liu, L. Gong, M.-L. Zhang, C. Cao, Y. He, The electronic properties and catalytic activity of precious metals adsorbed silicene for hydrogen evolution reaction and oxygen evolution reaction, *Appl. Surf. Sci.* 560 (2021) 150041. https://doi.org/10.1016/j.apsusc.2021.150041

[62] Z.-X. Guo, Y.-Y. Zhang, H. Xiang, X.-G. Gong, A. Oshiyama, Structural evolution and optoelectronic applications of multilayer silicene, *Phys. Rev. B* 92 (2015) 201413. https://doi.org/10.1103/PhysRevB.92.201413

[63] B. Huang, H.-X. Deng, H. Lee, M. Yoon, B.G. Sumpter, F. Liu, S.C. Smith, S.-H. Wei, Exceptional optoelectronic properties of hydrogenated bilayer silicene, *Phys. Rev. X* 4 (2014) 021029. https://doi.org/10.1103/PhysRevX.4.021029

[64] H. Emami-Nejad, A. Mir, Z. Lorestaniweiss, A. Farmani, R. Talebzadeh, First designing of a silicene-based optical MOSFET with outstanding performance, *Sci. Rep.* 13 (2023) 6563. https://doi.org/10.1038/s41598-023-33620-2

[65] M. Ezawa, Dirac theory and topological phases of silicon nanotube, *Europhys. Lett.* 98 (2012) 67001. https://doi.org/10.1209/0295-5075/98/67001

[66] M. Khalkhali, F. Khoeini, A. Rajabpour, Thermal transport in silicene nanotubes: Effects of length, grain boundary and strain, *Int. J. Heat Mass Transf.* 134 (2019) 503–510. https://doi.org/10.1016/j.ijheatmasstransfer.2019.01.074

[67] R. Chegel, Theoretical investigation of enhanced nonlinear optical properties of silicene and carbon nanotubes: Potential applications in infrared and ultraviolet optoelectronics, *J. Lumin.* 277 (2025) 120923. https://doi.org/10.1016/j.jlumin.2024.120923

[68] F. Salimian, D. Dideban, A silicene nanotube field effect transistor (SiNTFET) with an electrically induced gap and high value of Ion/Ioff, *ECS J. Solid State Sci. Technol.* 7 (2018) M1. https://doi.org/10.1149/2.0071802jss

[69] S. Cahangirov, M. Topsakal, E. Aktürk, H. Şahin, S. Ciraci, Two- and one-dimensional honeycomb structures of silicon and germanium, *Phys. Rev. Lett.* 102 (2009) 236804. https://doi.org/10.1103/PhysRevLett.102.236804

[70] S.M. Aghaei, M.M. Monshi, I. Torres, I. Calizo, Edge functionalization and doping effects on the stability, electronic and magnetic properties of silicene nanoribbons, *RSC Adv.* 6 (2016) 17046–17058. https://doi.org/10.1039/C5RA26107D

[71] Y.-L. Song, Y. Zhang, J.-M. Zhang, D.-B. Lu, Effects of the edge shape and the width on the structural and electronic properties of silicene nanoribbons, *Appl. Surf. Sci.* 256 (2010) 6313–6317. https://doi.org/10.1016/j.apsusc.2010.04.009

[72] A. Ramasamy, S.B.A, H.R.M. Raj, D.J. Thiruvadigal, DFT calculations on defect induced and doped ASiNR for sensing the COPD breath biomarker, *ECS J. Solid State Sci. Technol.* 13 (2024) 057001. https://doi.org/10.1149/2162-8777/ad40cf

[73] R. Akash, A.S. Balaji, K.J. Sivasankar, H.R. Mohanraj, D.J. Thiruvadigal, N doped armchair silicene nanoribbon as a promising biosensor for

detection of VOCs in healthcare monitoring: A first principles study, *Mater. Sci. Semicond. Process.* 189 (2025) 109302. https://doi.org/10.1016/j.mssp.2025.109302

[74] S. Showket, K.A. Shah, G.N. Dar, Pristine and modified silicene based volatile organic compound toxic gas sensor: A first principles study, *Phys. Scr.* 98 (2023) 085937. https://doi.org/10.1088/1402-4896/ace564

[75] Q. Li, H. Liu, Y. Tian, J. Guo, G. Chen, J.Y. Lee, Methylation detection and DNA sequencing based on adsorption of nucleobases on silicene nanoribbon, *J. Phys. Chem.* C 124 (2020) 10823–10831. https://doi.org/10.1021/acs.jpcc.0c01734

[76] S. Alesheikh, N. Shahtahmassebi, M.R. Roknabadi, R. Pilevar Shahri, Silicene nanoribbon as a new DNA sequencing device, *Phys. Lett. A* 382 (2018) 595–600. https://doi.org/10.1016/j.physleta.2017.12.010

[77] S.M. Aghaei, M.M. Monshi, I. Calizo, A theoretical study of gas adsorption on silicene nanoribbons and its application in a highly sensitive molecule sensor, *RSC Adv.* 6 (2016) 94417–94428. https://doi.org/10.1039/C6RA21293J

[78] G.K. Walia, D.K.K. Randhawa, Gas-sensing properties of armchair silicene nanoribbons towards carbon-based gases with single-molecule resolution, *Struct. Chem.* 29 (2018) 1893–1902. https://doi.org/10.1007/s11224-018-1170-9

[79] G.K. Walia, D.K.K. Randhawa, Adsorption and dissociation of sulfur-based toxic gas molecules on silicene nanoribbons: Aquest for high-performance gas sensors and catalysts, *J. Mol. Model.* 24 (2018) 94. https://doi.org/10.1007/s00894-018-3631-x

[80] G.K. Walia, D.K.K. Randhawa, First-principles investigation on defect-induced silicene nanoribbons—A superior media for sensing NH3, NO2 and NO gas molecules, *Surf. Sci.* 670 (2018) 33–43. https://doi.org/10.1016/j.susc.2017.12.013

[81] J. Kang, F. Wu, J. Li, Symmetry-dependent transport properties and magnetoresistance in zigzag silicene nanoribbons, *Appl. Phys. Lett.* 100 (2012) 233122. https://pubs.aip.org/aip/apl/article-abstract/100/23/233122/282138

[82] D. Zou, W. Zhao, C. Fang, B. Cui, D. Liu, The electronic transport properties of zigzag silicene nanoribbon slices with edge hydrogenation and oxidation, *Phys. Chem. Chem. Phys.* 18 (2016) 11513–11519. https://doi.org/10.1039/C6CP01159D

[83] F. Wan, X. Wang, Y. Guo, J. Zhang, Z. Wen, Y. Li, Role of line defect in the bandgap and transport properties of silicene nanoribbons, *Phys. Rev. B* 104 (2021) 195413. https://doi.org/10.1103/PhysRevB.104.195413

[84] H. Li, L. Wang, Q. Liu, J. Zheng, W.-N. Mei, Z. Gao, J. Shi, J. Lu, High performance silicene nanoribbon field effect transistors with current saturation, *Eur. Phys. J. B* 85 (2012) 274. https://doi.org/10.1140/epjb/e2012-30220-2

[85] M.W. Chuan, M.A. Riyadi, A. Hamzah, N.E. Alias, S. Mohamed Sultan, C.S. Lim, M.L.P. Tan, Impact of phonon scattering mechanisms on the performance of silicene nanoribbon field-effect transistors, *Results Phys.* 29 (2021) 104714. https://doi.org/10.1016/j.rinp.2021.104714

[86] A. Gooran-Shoorakchaly, A.A. Ahmadchally, S. Soleimani-Amiri, M. Gholipour, Design of a low-power short-channel electrostatically doped silicene nanoribbon FET, *IEEE Trans. Electron Devices* 68 (2021) 1956–1961. https://doi.org/10.1109/TED.2021.3058337

[87] M. Poljak, Impact of width scaling and parasitic series resistance on the performance of silicene nanoribbon MOSFETs, *IEEE Trans. Electron Devices* 67 (2020) 4705–4708. https://doi.org/10.1109/TED.2020.3017465

[88] C. Xu, G. Luo, Q. Liu, J. Zheng, Z. Zhang, S. Nagase, Z. Gao, J. Lu, Giant magnetoresistance in silicene nanoribbons, *Nanoscale* 4 (2012) 3111–3117. https://doi.org/10.1039/C2NR00037G

[89] M.A. Kharadi, I. Mahariq, Multi-functional spin-caloritronic device based on co-doped zigzag silicene nanoribbon heterojunction, *Results Phys.* 67 (2024) 108066. https://doi.org/10.1016/j.rinp.2024.108066

[90] D.-Q. Fang, S.-L.Zhang, H. Xu, Tuning the electronic and magnetic properties of zigzag silicene nanoribbons by edge hydrogenation and doping, *RSC Adv.* 3 (2013) 24075–24080. https://doi.org/10.1039/C3RA42720J

[91] G. Guo, Z. Xie, Y. Zhang, Y. Deng, J. Tang, Y. Mao, The ferromagnetic and half-metal properties of hydrogen adatoms, fluorine adatoms and boron adatoms adsorbed at edges of zigzag silicene nanoribbon, *Phys. E Low Dimens. Syst. Nanostruct.* 116 (2020) 113733. https://doi.org/10.1016/j.physe.2019.113733

[92] X.Q. Deng, Z.H. Zhang, G.P. Tang, Z.Q. Fan, L. Sun, C.X. Li, H.L. Zhang, The design of bipolar spin filtering junction in zigzag silicene nanoribbons, *Org. Electron.* 37 (2016) 245–251. https://doi.org/10.1016/j.orgel.2016.06.037

[93] S. Kharwar, S. Singh, N.K. Jaiswal, Structural & electronic properties of zigzag silicene nanoribbons with symmetric/asymmetric edge passivations via fluorine and hydrogen, *Phys. B Condens. Matter* 615 (2021) 413072. https://doi.org/10.1016/j.physb.2021.413072

[94] Y. Mao, C. Tang, G. Guo, J. Yuan, J. Zhong, Evolution of the electronic and magnetic properties of zigzag silicene nanoribbon used for hydrogen storage material, *Int. J. Hydrog. Energy* 42 (2017) 27184–27205. https://doi.org/10.1016/j.ijhydene.2017.09.088

[95] G. Guo, Y. Mao, J. Zhong, J. Yuan, H. Zhao, Design lithium storage materials by lithium adatoms adsorption at the edges of zigzag silicene nanoribbon: A first principle study, *Appl. Surf. Sci.* 406 (2017) 161–169. https://doi.org/10.1016/j.apsusc.2017.02.053

[96] L. Xu, X.-F. Wang, L. Zhou, Z.-Y. Yang, Adsorption of Ti atoms on zigzag silicene nanoribbons: Influence on electric, magnetic, and thermoelectric properties, *J. Phys. Appl. Phys.* 48 (2015) 215306. https://doi.org/10.1088/0022-3727/48/21/215306

[97] M.A. Kharadi, G.-F.A. Malik, K.A. Shah, F.A. Khanday, Performance analysis of functionalized silicene nanoribbon-based photodetector, *Int. J. Numer. Model. Electron. Netw. Devices Fields* 34 (2021) e2809. https://doi.org/10.1002/jnm.2809

[98] Z.A. Azad, A. Shokri, M.S.A. Khezrabad, First-principle study on the quantum transport in an armchair silicene nanoribbon-based p–n diode,

Mater. Chem. Phys. 257 (2021) 123483. https://doi.org/10.1016/j.matchemphys.2020.123483

[99] Mubashir A. Kharadi, Gul Faroz A. Malik, Farooq A. Khanday, Khurshed A. Shah, Sparsh Mittal, Brajesh Kumar Kaushik, Silicene: From material to device applications, *ECS J. Solid State Sci. Technol.* 9 (2020) 115031. https://iopscience.iop.org/article/10.1149/2162-8777/abd09a/meta

[100] Shazia Showket, Khurshed A. Shah, G. N. Dar, Pristine and modified silicene based volatile organic compound toxic gas sensor: A first principles study, *Phys. Scrip.* 98 (2023) 085937. https://iopscience.iop.org/article/10.1088/1402-4896/ace564/meta

Chapter 7

Perovskites

Revolutionary materials for next-generation devices

S. Gomathi, J. S. Nirmalram, M. Nidhishree, Roweena Agnes, S. Subash Chandra Bose, and Muhammdh Sahaafi

7.1 INTRODUCTION

7.1.1 History

Perovskite materials[1,2] which were initially identified in 1839 as calcium titanium oxide minerals[3] have developed into a very adaptable family of substances with a wide range of uses. Perovskites generally have the composition ABX_3 ('A' and 'B'–cations of various sizes; 'X'–an oxygen anion).

Perovskite, a mineral having a crystal structure that has been thoroughly investigated for more than 60 years, has proven essential to many scientific disciplines due to its application in high-temperature superconductivity, enormous magneto-resistance, and ferroelectricity.[4] Research indicates that perovskite significantly influences Earth's lower mantle, and argon can dissolve in $MgSiO_3$ perovskite, thereby suggesting the presence of argon in the lower mantle throughout Earth's early history.[5] Perovskite's crystallization history in various geological contexts showed that in kimberlite rocks, depending on the makeup of the magma, perovskite grains crystallize after other macrocrystal phases including olivine and spinel.[6] The emplacement history of carbonatite and silicate rocks in the Oka Carbonatite Complex in Canada has also been illuminated by studies revealing a multi-stage petrogenetic history documented by perovskite and apatite minerals.[7] Perovskite has become a focal point in the solar energy sector because of its use in photovoltaic devices. Beginning in the early 2010s, halide perovskite photovoltaics have advanced quickly, resulting in a marked growth in energy conversion capability over time.[8] Understanding the interfacial mechanism at the charge extraction zone—perovskite boundary is substantial in augmenting the efficacy of hybrid perovskite solar cell (PeSC) involving organic and inorganic materials through heterojunction modifications, as evidenced by recent studies.[9] Furthermore, recent advancements in the field of perovskite materials have introduced new structures like the M_3AB-type hybrid organic–inorganic anti-perovskite ferroelectrics, thus broadening the research beyond conventional mineralogy.[10] Perovskite's extensive

DOI: 10.1201/9781003512899-7

history and wide range of applications highlight its importance across scientific fields, including geology, materials science, and renewable energy. The literature on the history of perovskites reflects a complex and intricate research trajectory that is always evolving in reaction to novel discoveries and technological advancements.

Barium titanate ($BaTiO_3$) perovskite minerals were found to exhibit ferroelectricity[11] at the beginning of the 20th century. This means that they possess a reversible spontaneous dielectric polarization under an applied electric field. Due to their beneficial dielectric properties, perovskites were used in capacitors and other electronic devices as a result of this discovery.[12] In the 1980s, cuprate-based perovskites, notably yttrium barium copper oxide[13] ($YBa_2Cu_3O_7$), were identified as high-temperature superconductors which created new opportunities in magnetic levitation, and energy transmission technologies. A team led by Tsutomu Miyasaka made significant strides in PeSCs[14–17] in 2009, demonstrating the pioneering use of perovskites in solar energy utilization. With their low efficiency at first, notable advancements made by 2012 significantly raised awareness of their potential. Since then, the efficacy of these PeSCs has augmented dramatically, from 3.8% to 25% in a decade making them a viable option for flexible, affordable, and effective solar power technology.

7.2 CHARACTERIZATION

7.2.1 Raman spectroscopy

It is a key tool for describing perovskite materials as it sheds light on their vibrational and structural characteristics. Through the detection of bond lengths and angles, vibrational mode observation for atomic motions, and phase identification through distinct Raman fingerprints, Raman spectroscopy facilitates structural study. In addition, flaws, impurities, and dopants are detected using Raman spectroscopy, which identifies alterations connected to stability and degradation under environmental stress. This method effectively characterizes perovskite materials, aiding in crystal structure evaluation, dimensionality prediction,[18] octahedral connectivity, and lattice mode symmetry analysis.[19] Its spectra reveal vibrational modes, identifying chemical bonds and structural elements. Structural analysis of lead halide perovskites like $CsPbI_3$[20,21] highlights Raman's role in studying spectra–structure relationships in $APbI_3$. Advanced studies of methylammonium lead iodide perovskites[22] underscore the need for precise characterization for solar cell stability. The optical impact of dopants, such as Eu^{3+} in $CaTiO_3$ and Cr^{3+} in TiO_2, has been observed.[23] This adaptable technique expands our knowledge of perovskites for use in solar cells, light-emitting diodes (LEDs), and catalysis and is frequently used in conjunction with other analytical techniques.

7.2.2 Powder XRD

Doping lanthanide ions, like Ce^{3+} and Yb^{3+}, into perovskite nanocrystals (NCs) enhances optical properties while retaining the tetragonal P4mm crystal structure, with lattice contraction from 3.94 to 3.87 Å as atomic number increases. In $LaMn_{1-x}Co_xO_3$ perovskites, X-ray diffraction (XRD) reveals[24] larger particles at higher cobalt substitution, with surface enrichment in higher oxidation states of B cations, indicating potential for tunable electrocatalytic properties while maintaining homogeneous bulk composition. The synthesis of type $ABB'O_3$ (the A-site elements include La, Pr, and Nd, the B-site elements consist of Fe and Mn, while the B′-site elements are Pd and Pt) *via* $Pd(NO_3)_2$ and $PdCl_2$ show pure perovskite formation with minor impurities like A_2BO_3 or A_2BO_4, thereby confirming structure *via* XRD. In $MAPbBr_3$, crystallized from $PbBr_2$ and MABr, synchrotron XRD reveals coherent scattering parameters throughout Pb, Br, C, N, and H atoms, with symmetry shifts upon cooling, specifically the Imma phase at 150K, also seen in related perovskites like $SrRuO_3$. $CsPbBr_3$ exhibits phase transitions from orthorhombic to cubic at elevated temperatures, confirmed by Rietveld refinement and unit cell expansion. $APbI_3$ perovskites (A = Na, Cs, Li) show hexagonal phase and crystallinity with $NaPbI_3$ and $LiPbI_3$ favoring the (002) plane and $CsPbI_3$ the (003) plane, as per XRD. The $AgNb_7O_{18}$ photocatalyst with no impurities displays an orthorhombic perovskite-like structure. Mechanochemically fabricated double perovskites, such as $Cs_2AgBiBr_6$ and $MA_2TlBiBr_6$, confirm purity, although $Cs_2AgSbBr_6$ shows $Cs_3Sb_2Br_9$ impurities and unidentified phases. XRD of $MAPbI_3$ films reveals phase shifts from a yellow intermediate to black perovskite upon annealing and X-ray exposure, suggesting possible orientation changes. The modified solid-state synthesis of $KNbO_3$ perovskite shows phase evolution from a metastable phase at 500°C to the perovskite-type structure at 550°C, as observed through temperature-dependent XRD analysis.

7.3 APPLICATIONS OF PEROVSKITE MATERIALS

Optoelectronic applications have been transformed by perovskite materials because of their exceptional structural plasticity, which allows for tunable electronic and optical properties that may be adjusted for a variety of technologies. Enhanced carrier mobility, ideal absorption characteristics, and economical processing are just a few of the many benefits that perovskites provide for anything from high-efficiency solar cells to sophisticated photodetectors, sensors, lasers, and LEDs. Recent developments have further enhanced the stability and performance of perovskites, thus making them a potential material for next-generation devices that require scalability and great efficiency. This section explores various applications of perovskite materials, focusing on important developments and commercialization opportunities in optoelectronic and photonic technology.

7.3.1 Perovskite-based solar cells

Perovskite materials are groundbreaking in optoelectronics, especially for photovoltaic applications, because of their flexible bandgap, significant absorption coefficient, impressive charge transport mobility, and wide absorption range. The photovoltaic conversion behavior of PeSC has significantly improved (21.7% in one decade).[25,26] Various types of perovskite materials exhibit different power conversion efficiency (PCE) advancements. For example, oxide perovskites, such as $CaTiO_3$ and its self-deficient forms ($Ca_{1-\alpha}Ti_1O_3$, $CaTi_{1-\beta}O_3$, and $CaTiO_{3-\gamma}$), were dip-coated onto indium tin oxide (ITO) substrates, with XRD and scanning electron microscopy (SEM) analyses confirming their structure. Photocurrent measurements showed that O-deficient $CaTiO_3$ had a notably higher photocurrent under ultraviolet (UV) light, while Ti and Ca-deficient films have lower values. Self-deficient $CaTiO_3$ also exhibited optimal UV responsivity and detectivity, with density functional theory (DFT) calculations supporting these results, suggesting improved device performance. Incident Photon-to-Current Efficiency (IPCE) reached 25%–28% in the UV range between 280 and 400 nm.[27]

Halide perovskite materials, particularly cesium lead triiodide ($CsPbI_3$), are notable for their photovoltaic characteristics and enhanced PCE. $CsPbI_3$, with favorable bandgap (~1.75 eV) and thermal stability, is promising for tandem solar cells, although high-efficiency achievement is complex. This study presents a formamidine acetate-assisted procedure to create homogeneous phase γ-$CsPbI_3$ films, enhancing crystallinity and interfacial contact, achieving over 18% PCE and sustaining 95% efficiency after 30 days.[28] For $CsPbI_2Br$ cells, thermal stability is offset by lower PCEs due to reduced open-circuit voltages (VOC). The study identifies a quasi-Fermi level splitting (QFLS) mismatch, particularly high in p–i–n cells (179 meV) compared to organic–inorganic perovskites (11 meV). $CsPbI_2Br$ films, with a 1.9 eV bandgap, display low defect density and a potential efficiency of 20.8%, with recommended strategies to address interface misalignment and improve VOC.[29]

Copper (I) thiocyanate is a potent hole extraction substance for perovskite photovoltaic cells, along $MAPbI_3$. Spin-coating CuSCN dissolved in $(C_2H_5)_2S$ requires careful post-annealing to avoid $MAPbI_3$ degradation, but increasing the molar ratio of CH_3NH_3I in $MAPbI_3$ helps mitigate surface damage. This method produced a stable CuSCN/$MAPbI_3$ (excess MAI)/TiO_2 cell with 17% efficiency and improved damp heat stability over spiro-OMeTAD.[30] Lead-free alternatives like tin and bismuth halides are also advancing. Tin-based $FASnI_3$, enhanced with trimethylthiourea, improved film quality and gained 14% PCE, resulting in enhanced moisture resilience.

Separately, $(NH_4)_3Bi_2I_9$ perovskites (2.1 eV bandgap) achieved 0.6% PCE *via* solvent engineering.[31]

$CsBiSCl_2$, synthesized with $DMABiS_2$ intermediates, reached a 10.38% PCE with minimal efficiency drop after 150 days in air, showing eco-friendly promise.[32] Cesium-doped $FASnI_3$ perovskites, optimized *via* Solar Cell Capacitance Simulator (SCAPS) simulation with ZnOS electron transport layer (ETL), achieved over 20% simulated PCE, with $FA_{0.75}Cs_{0.25}SnI_3$ showing particular stability and efficiency.[33] Cs_2BiAgI_6 is also being explored for its cost-effectiveness and stability, with Solar Cell Capacitance Simulation in 1 Dimension (SCAPS-1D) simulations optimizing absorber and ETL parameters, yielding 23.71% PCE in phenyl-C61-butyric acid methyl ester (PCBM)-based cells, suggesting potential for commercial applications.[34]

Lead-free double perovskites like Cs_2SnI_6 and $Cs_2InBiBr_6$ are promising eco-friendly solar cell materials. SCAPS-1D simulations optimized these materials, showing 23.64% simulated PCE for Cs_2SnI_6 having a GO ETL and Cu_2O HTL, and similar results for $Cs_2InBiBr_6$ in an $FTO/TiO_2/Cs_2InBiBr_6/Cu_2O/Au$ setup, with both materials demonstrating excellent stability potential.[35] In two-dimensional Pb–Sn hybrid PeSC (iso-$BA_2MA_4(Pb_{0.7}Sn_{1-0.7})_5I_{16}$), adjusting the ratio of Pb/Sn in (n) = 5 resulted in 18.06% efficiency. Higher Pb content promoted vertical 2D phase growth, improving stability and performance by reducing lattice strain and enhancing organic-inorganic interactions.[36] While perovskites have excellent optoelectronic properties, challenges remain in optimizing absorber materials, device fabrication, film morphology, additives, crystalline nature, composition, and interface properties. Addressing these will be important to advancing commercial applications.

7.3.2 Perovskite-based laser

The successful growth and characterization of a Tm^{3+} and Ho^{3+} -co-doped $GdScO_3$ perovskite crystal[37]using the Czochralski method reported in the literature contain 3.76 at.% Tm and 0.35 at.% Ho. This crystal exhibited broadband emission near 2 μm, with energy transfer parameters of (P28 and P71= 1.30×10^{-22} and 0.99×10^{-23} $cm^3/\mu s$ respectively) resulting in 3.5 ms thermal lifespan. The CF splitting and spectral line broadening mechanisms, reported a steady-state laser output, slope efficiency, and threshold of 1.16 W, around 2.1 μm, 50.5% and 184 mW, respectively. These observations reveal that Tm/Ho-doped gadolinium scandate is the optimal choice for tunable wavelength mode-locked lasers operating >2 μm. Saturable absorber (SA) made from tapered fiber and Ba_2LaTaO_6 for generating ultrashort pulses in fiber lasers was synthesized through a high-temperature solid-state process.[38] This absorber features 2.66 MW/cm^2 saturation intensity and 24% modulation depth. By manipulating intra-cavity chromatic dispersion in conjunction with a continuous wave laser, various pulse types can be generated in a passive fiber laser with mode-locked, marking the first application of Ba_2LaTaO_6 and highlighting its potential in ultrafast photonics. Lead halide perovskite NCs[39], specifically $CsPbBr_3$, have been used to develop a simple vertical-cavity surface-emitting laser (VCSEL) that emits at 527 nm and features a low lasing threshold. The $CsPbBr_3$ NCs were synthesized through a recrystallization method. Time-resolved photoluminescence studies revealed carrier dynamics, including gain-switching and a rapid pulse decay of approximately 10 ps, contributing to the understanding of $CsPbBr_3$ NC VCSELs for optoelectronic applications. The TiO_2 microspheres combined with $Cs_{0.5}Rb_{0.5}PbBr_3$ quantum dots[40] (QDs) can achieve microcavity lasing which exhibits excellent photostability and quantum yield, outperforming other halide QDs. Lasing is initiated when the 7.98 kW/cm^2 threshold is exceeded, with significant amplification observed using an Argon-ion laser at 488 nm, resulting in a 1807 quality factor under ambient conditions. Exceptionally, the QDs retain photostability under continuous laser power of 20.03 kW/cm^2 for 90 minutes, highlighting the potential of Rb-doped halide QDs for tunable, stable microlaser applications.

The effects of substituting lead with copper or tin in hybrid perovskites in X-ray imaging were reported and resulted in an increased bandgap of 3.90 eV for copper and 3.44 eV for tin, along with enhanced photoluminescence decay times. However, both alternatives exhibit radioluminescence at room temperature, and both experience significant thermal quenching, leading to reduced light yields. The thermal evaporation technique was reported for producing high-performance random lasers from perovskite thin films enhanced with triphenylphosphine oxide (TPPO), achieving an optical gain five times greater than pristine films, along with a lower lasing threshold and high polarization. These findings present a scalable approach for high-output perovskite materials and related devices, supporting cutting-edge speckle-free imaging systems.[41] The perovskites of the type, non-lead $Cs_3Cu_2I_5$ for laser applications were created *via* dual-feed co-evaporation as thin films achieve a photoluminescence quantum yield of 50.62%. Pe-VCSELs utilizing these films demonstrate laser emission at 440 nm with 1.12 μJ/cm^2 threshold and the dual-frequency laser emits at 458.8 and 505.6 nm. The research also introduces a lead-free phosphor–sapphire composite (PSC) containing Rb_2SnCl_6, enhancing thermal resistance and producing chartreuse light emission (570 nm). When a blue laser is used in conjunction with this, PSC generates amber-white light having a correlated chromatic temperature (4129 K), highlighting the prospects of Pb-free perovskite substances for high-power laser-driven lighting.[42]

In addition, the Pb-free double perovskites, particularly $Cs_2AgBiBr_6$, are used for ultrafast photonics. $Cs_2AgBiBr_6$ is environmentally friendly, stable, and demonstrates strong nonlinear absorption. The researchers synthesized $Cs_2AgBiBr_6$ NCs and assessed their performance as a SA in Er-doped fiber lasers, achieving a saturation intensity of 0.75 MW/cm^2 and 50 mW threshold power for stable mode-locking. This fiber laser-generated pulses as short as 452 fs, and the $Cs_2AgBiBr_6$ SA exhibited excellent long-term stability, remaining effective after 10 months. These findings underscore $Cs_2AgBiBr_6$'s potential in ultrafast photonics.[43] The study reported that two-photon induced amplification of spontaneous emission in 2D Ruddlesden-Popper perovskite (RPP) flakes at ambient temperature. $(BA)_2(MA)_{n-1}Pb_nI_{3n+1}$ high-quality flakes were created, showcasing tunable emission from 527 to 680 nm. Two-photon pumped amplified spontaneous emission (ASE) was observed in flakes with n values of 3, 4, and 5, exhibiting lower thresholds than many semiconductor nanostructures, thus highlighting the strong performance of RPPs. The study also explored pump-wavelength-dependent ASE behaviors, revealing the capability of near-infrared laser light to convert to visible light *via* frequency upconversion in RPPs, opening avenues for wavelength-tunable frequency upconversion.[44] Lastly, the exploration of double junction (DJ) configured quasi-2D perovskites as laser gain media indicates that isolated carrier recombination can hinder ASE. Treatment with Cremophor EL improves the distribution of large n domains, increasing radiation-state density and carrier interaction, which lowers the ASE threshold from 11.7 to 4.8 $\mu J/cm^2$, boosts the optical gain coefficient from 775 to 1559/cm, and enhances the polarization degree from 0.59 to 0.98. These results emphasize the importance of optimizing carrier co-radiation recombination for optimizing the performance of DJ structured perovskite lasers with quasi-2D structures.[45]

7.3.3 Perovskite-based LEDs

Perovskite LEDs have become viable candidates in optoelectronics, combining high efficiency with cost-effective fabrication methods and high electroluminescence efficiency achieved through solution processing.[46] PeLEDs show significant potential in applications such as optically pumped lasers and LEDs.[47] However, limitations like device degradation and challenges in maintaining high efficiency under large current densities have hindered progress toward electrically driven perovskite laser diodes.[48] To enhance PeLED efficiency and stability, various strategies have been investigated. For example, lithium halide passivation has been effective in reducing trap densities and suppressing ion migration in perovskite films, thereby improving blue and red PeLED efficiency.[49] In addition, interfacial charge-confinement structures have been introduced to enhance carrier trapping and exciton formation in phosphorescent emitters, enabling efficient white light emission for displays and lighting applications.[50] Lead-free perovskite derivatives have also been explored to address stability and efficiency concerns. Studies on lead-free organic-perovskite hybrid quantum wells have demonstrated efficient and consistent light emission, highlighting their promise in amber white LEDs,[51] while non-lead double perovskite halide NCs exhibit improved stability and efficiency in light-emitting applications.[52] In terms of material properties, $CsPbBr_3$ NCs doped with nickel have shown enhanced optical stability, enhancing PeLED performance.[53] Additionally, the use of trapped optical modes in white PeLEDs has led to efficiencies exceeding 12%, underlining the value of innovative approaches in achieving high-performance devices.[54] For green LEDs, large CsPbBr NCs have been utilized to improve color purity and efficiency, further demonstrating the role of material properties in device performance.[55] Thus, these studies emphasize the ongoing efforts to enhance PeLED efficiency and stability *via* innovations in materials, device configurations, and defect passivation. Although challenges remain for electrically driven perovskite laser diodes, the reviewed research highlights promising avenues for future advances in perovskite optoelectronics.

7.3.4 Perovskite-based photodetector

Recent advancements in optoelectronics have led to increased attention to perovskite materials for photodetector applications. Perovskites outperform silicon due to their higher absorption coefficient, lower production costs, configurable bandgaps, faster responsiveness, detectivity, and superior photoelectric performance. Perovskite-based photodetectors are also well-known for their affordable, flexible, and adaptable photodetection engineering. In a photodetector, the optical signals are converted to electrical impulses which are employed in several domains, including military surveillance, bionics, optical communication, and electronic image processing. Perovskite photodetectors face challenges due to their low quantum efficiency and high energy consumption; however, advancements in synthesis methods and innovative device designs have significantly enhanced their photodetection mechanisms.[56] A study by Neda F. highlighted that various perovskites, including ZnO nanorods coated with $BaTiO_3$ nanoparticles, demonstrate enhanced UV photodetection characteristics. Nazari *et al.* used a novel vapor thermal synthesis technique. Under 365 nm UV light, a $BaTiO_3$ decorated ZnO nanorod photodetector exhibited short decay times (46–0.96 seconds), enhanced responsiveness (57–135), and quick rising times[57] (2–0.75 seconds). DFT analysis showed that the bandgaps for Ca-doped $RaTiO_3$ were 1.96 eV, 1.92 eV, and 1.87 eV for different $Ra_{0.875}Ca_{0.125}TiO_3$, $Ra_{0.75}Ca_{0.25}TiO_3$, and $Ra_{0.625}Ca_{0.375}TiO_3$ concentrations, respectively. This suggests promising potential for optoelectronic applications,[58] particularly in examining the response of Ca inclusion on the optical and electrical characteristics of $RaTiO_3$ for energy applications.

High-performance 2D perovskite $Pb_2Nb_3O_{10}$ photodetectors, fabricated *via* calcination followed by liquid exfoliation, exhibit high responsivity, detectivity, and fast speed and the values are (2.8 A/W), (1.1×10^{12} Jones) and (0.2/1.2 ms) respectively, and under 350 nm radiation,[59] 2D Dion-Jacobson perovskite oxides $Ca_2Nb_{3-x}Ta_xO_{10}$; where (x is 0, 0.5, 1, 1.5) exhibits superior execution by having photocurrent (31.4 μA), a greater ratio (on-off =5.6×10^4) and an enriched responsivity (469.5 A/W) at optimized condition, reaching 7000 fold greater than pristine.[60] Incorporating 2D $Sr_2Nb_3O_{10}$ nanosheets and N_2-doped graphene QDs, reveals a remarkable change of photocurrent (0.47 μA), on–off (2.2×10^4), and sensitivity speed[61] (τ_{decay}) at 67.3 ms was developed. Hybrid organic–inorganic perovskite category is noticed widely for its strong optical absorption capabilities, excellent carrier mobility and extended carrier diffusion distance.[62] D_xMAPbI_3 single crystals were grown by varying deuterium levels (x = 0, 0.15, 0.99). The $D_{0.15}MAPbI_3$ variant exhibited an excellent external quantum efficiency (EQE), demonstrated improved thermal stability with reduced photocurrent degradation[63] compared to pure samples, and showed greater responsivity (2.8 A/W), efficient detectivity (1.1×10^{12}Jones), and good response period (0.2/1.2 ms) under 350 nm radiation.

Lead-based perovskites are a suitable source for optical and electronic devices as a result of their notable features including elevated absorption coefficient, a long carrier duration, a direct bandgap, fewer defect states, etc. The increased (5.3×10^{13}Jones) detectivity and broad unidirectional sensitivity range (124 dB) of Mg implanted $NiO/CH_3NH_3PbI_3/C_{60}$/bathocuproine/Cu is similar to that in Si photodetectors (120 dB).[64] A wide-area Si nanopillar formation was created by the organic–inorganic halide family of perovskite, $MAPbI_3$ thin films with nanospheres made using nanosphere lithography and dry etching. This yielded 11.2 A/W responsivity and 7.3×10^{10} Jones of detectivity at 450 nm, which is better than the thin films made without nanostructures.

Halide perovskites based on cesium show promising optical properties for enhanced photodetector performance. A method was developed by Zhou *et al.* for fabricating monolayer $CsPbBr_3/Cl_3$ NC films by using a spin coating technique with a two-stage doping of halide resulted in an on-off value (10^{-6}), responsivity (0.23 A/W) and open-circuit potential (1.6 V) in the near-UV spectrum.[65]

Despite their exceptional optoelectronic properties, the consistency and toxicity in lead-based perovskite substances hinder their commercial use, prompting the innovation of more stable cum sustainable non-lead alternatives. $CsSnBr_3$ QDs are a promising alternative for UV photodetectors[66] because of their low defect density and 1.27×10^{11} Jones detectivity at 365 nm. Lead-free perovskites are preferred because they are non-toxic, however they frequently perform poorly. The Ag_2BiI_5 lead-free perovskite film, which was created *via* a solution technique, has excellent performance traits. Compared to previous lead-free options, its $FTO/SnO_2/Ag_2BiI_5$/carbon structure[67] achieves a linear sensitivity range of 138 dB and 5.3×10^{12} Jones of detection limits.

Double perovskite oxides of the formula $A_2BB'O_6$, known for their strong structural stability and compositional adaptability, exhibit improved photocurrent capacity and sensitivity ranges between 365 and 760 nm. Notably, $Ba_2Bi_{1.5}Nb_{0.5}O_6$ achieves a photocurrent and light sensitivity of 59.2 μA/cm^2 and 78.8 μA/W below 365 nm radiation, respectively, and it was found to be 60.5 times greater than Ba_2BiNbO_6 based photodetectors.[68]

Halide lead-free double perovskites are notable for their excellent stability and potential in photodetection, with currently low-performance levels. The $Cs_2AgBiBr_6$:Na^+,Ce^{3+} film demonstrates high detection limits of 1.34×10^{12} and 1.18×10^{12} J at 460 and 370 nm, respectively, outperforming other lead-free halide double perovskites.[69]

When combined with the $CsPbI_3$ perovskite photodetector, the $Cs_2NaYb_{0.98}Tm_{0.02}Cl_6$ double perovskites with capable upconversion emission provide effective visible and near-infrared detection. The detection limit for 980 nm light can attain up to 1.26×10^{11} Jones, the maximum value for self-powered near infrared photodetectors.[70]

Dion–Jacobson phase has remarkable importance because of its exceptional optical and electrical properties present in the layered perovskite class. (HIS)(DMA)Pb_2Br_7 is a recently developed class of Donor Junction-type 2D perovskites. Its perovskite cavities have distinct structural distortion to make it a relatively larger secondary DMA^+ cation, establishing a strong phase stability that results in high detectivity of 1.2×10^{13} Jones, responsivity of 18.1 A/W, polarization-sensitive detection,[71] and a path toward photodetection applications.

These kinds of perovskites, along with their remarkable optical features, particularly their photodetection capabilities are useful. There are still several practical issues and obstacles with perovskite-based photodetectors, particularly with regard to their stability in the exposed atmosphere, regardless of their widespread use. Therefore, the controlled synthesis process, repeatability, and surface passivation are all appropriate methods to accurately correct the degradations. By taking these things into account and changing the processes to produce high-quality and stable material, the potential of perovskites in photodetector applications will be enhanced.

7.3.5 Perovskite-based sensors

Perovskites have been used in chemosensing technology in detecting numerous analytes like heavy metal ions, gases, solvents, drugs, and biomolecules with high sensitivity to the changes in optical/ electrochemical signals and rapid real-time detection by using their high fluorescence, redox capability and quantum confinement nature.

7.3.5.1 Electrochemical sensors for metal and biomolecule detection

It is a rapid, affordable, compact, and user-friendly device for monitoring environmental pollution offering good selectivity.[72,73] It offers good selectivity and high sensitivity in comparison to alternative methods. Anodic stripping voltammetry technique, an electrochemical method, is used commonly to detect metal ions.[74]

Inorganic perovskite NCs (ABO_3 or A_2BO_4 type structures) are used in electrochemical sensing applications, with alkaline-earth metal perovskites (Ca, Mg, Ba, Sr) offering affordability, reactivity, and enhanced stability.[75] Metal titanates add non-toxicity and conductivity.[76] Lanthanide perovskites, like $NdFeO_3$, enable detection of biomolecules such as dopamine, uric acid,[77] hydroquinone, catechol,[78] amlodipine, and ascorbic acid[79] as reliable chromatography alternatives. A La–Co nitrate–hydrotalcite composite detects flutamide[80] with elevated sensitivity (0.7571 μAμ/M cm^2, 0.002 μM detection limit). $Pr_{1-x}Sr_xCoO_{3-\delta}$ compositions ($x = 0–1$) allow selective, stable p-phenylenediamine detection[81] in hair dye. N-rGO/$SrZrO_3$ cubes detect hydroquinone[82] at 0.61 μM, while $SrCeO_3$ nanoparticles selectively sense nifedipine in urine, water, and Neu differentiation factor (NDF) tablets,[83] even amide nitro-containing drugs and pollutants. $CeBiO_x$ nanofiber electrodes provide high-sensitivity acetaminophen sensing (360, 350 μA m/Mcm^2) with substandard detection limits.[84]

7.3.5.2 Perovskite-based sensing in aqueous media

Coating perovskite surfaces with hydrophobic materials (alkyl phosphonic acid ω-ammonium chlorides,[85] metal-organic frameworks (MOFs),[86] polyethylene glycol (PEG),[87] poly (vinyl pyrrolidone),[88] stearic acid,[89] acid-base additives[90] or capping with PCBEG and PEG-60 fullerenes[91] improves their stability in aqueous media, enabling their use as probes in sensing applications within water-based environments. Phenethylammonium stabilized 2D PEA_2PbI_4 (where PEA is phenethylammonium) perovskite NCs,[92] $CH_3NH_3PbBr_3$ QDs encapsulated MOF-5 matrix and dual ligand capped $CH_3NH_3PbBr_3$ QDs are, few examples, made of perovskite-based sensors working under aqueous environment.

7.3.5.3 Gas sensors

The operation of gas-sensing perovskite devices primarily relies on (i) electron transfer and (ii) adsorption mechanisms. Analyte gases (e.g., NO_2, CO, O_3, H_2S, and H_2) interact with electrons under applied voltage, forming reactive oxygen species that reduce to oxygen, which adsorbs on the perovskite layer, changing electrochemical signals for sensing.

Organometal halide perovskite films were developed for O_2 detection *via* resistive response using a gold reference electrode.[93] A $CH_3NH_3PbI_{3-x}Cl_x$ perovskite-coated Pt electrode enables self-powered ozone sensing, achieving high sensitivity at 2500 ppb after 60 minutes *via* electrochemical detection exhibiting long-term stability and compatibility.[94] A sensor probe *via* fabrication of a thin layer of Ag-doped $CaCu_3Ti_4O_{12}$ was used for increasing hydrogen disulfide gas detection sensitivity.[95] The ((La, Sr) $(Cr,Fe)O_{3-\delta})$/YSZ/Pt sensor exhibits hydrogen selectivity, without influence of other gases. Cerium-doped[96] La_2CuO_4 (between 50 and 600 ppm at 350°C), a bismuth ferrite/Au sensor[97] (30 ppm at 350 °C, 150-day stability), and a TiO_2/(La, Sr, Co, Ni, O) heterojunction (400 ppm at 200 °C) were developed for enhanced CO detection. The $CH_3NH_3PbI_{3-x}(SCN)_x$ $CaCu_3Ti_4O_{12}$ resistive chemical sensor detects acetone and nitrogen dioxide at ambient temperature, with limits of 200 ppb and 20 ppm, respectively, while modified $CH_3NH_3PbBr_3$ films detect NO_2 down to 0.1 ppm. A cost-effective $LaFeO_3$ hollow micro spindle sensor[98] with an Ag reference electrode achieved a 5 ppm NO_2 detection limit at 155°C, and PdO nanoparticle enhancement improved its sensitivity, stability, and selectivity. Additionally, an yttria-stabilized zirconia-$CoTiO_3$ electrode achieved 500 ppb NO_2 detection limit at 650°C, thereby maintaining stability[99] for 20 days. A Cu-doped $NaBiTiO_3$ perovskite sensor detected NO_2 at 2.25 μA for concentration levels (0.15–500 ppm) at 500°C.

7.3.5.4 Optical sensors

Perovskite NCs gain their optoelectronic properties like long charge carrier mobility parameters, exceptionally narrow emission spectra and elevated photoluminescence quantum efficiency by the quantum confinement, thus obtaining high sensitivity, enhanced detection limits, improved signal-to-noise ratio, and energy efficiency. These sensors interact with metal ions through the fluorescent quenching mechanisms or fluorescent enhancement mechanisms like cation exchange, anion exchange, electron transfer, and Förster resonance energy transfer mechanisms.[100]

7.3.5.5 Water sensors

Various perovskite-based sensors have been developed for ultra-sensitive water detection.[101–106] $CsPbBr_3$ perovskites on paper substrates detect humidity in pharmaceutical samples, while dual-emission $CsPbBr_3$@DMT-NH_2 QDs provide visual water detection in oils (detection limit: 0.006%). Lead-free $Cs_2InBr_5.H_2O$ and $Cs_3Sb_2Br_9$ NCs serve as effective trace water sensors in solvents, with 0.033% and 0.02 ppm detection limits, correspondingly. The eco-friendly PEA_6SnI_8 crystal offers reversible humidity sensing *via* luminescent chromism, underscoring perovskites' potential in environmental and analytical applications. Moreover, a low-voltage (20 mV) impedance sensor using $CsPbBr_3$ nanoparticles exhibits fast response and stability, showcasing strong potential for real-time health monitoring.

7.3.5.6 Radiation sensors

Halide lead perovskites show great potential for detecting ionizing radiation detection, including X-ray detection and γ-spectroscopy, owing to their strong radiation resistance, defect tolerance of defects, high mobility-charge carrier duration, adjustable bandgap and cost-efficient crystallization, with some challenges. Alloying $MAPbBr_3$ with n-type $MAPbCl_3$ induces an n-type transition, with $MAPbBr_{2.94}Cl_{0.06}$ showing the lowest charge carrier concentration and increased mobility-lifetime[107] product ((1.2–1.8) × 10^{-2} cm^2/V). $FAPbI_3$ single crystals for γ-ray detection achieve a maximum mobility lifespan product (1/100 cm^2/V) and respond well to gamma-ray photons (59.6 keV; 241Am). To stabilize the thermodynamically unstable black phase, and enhance stability and performance, a mixture of single crystalline halide perovskite with the composition of $Cs_{0-0.1}FA_{1-(0-0.1)}PbI_{3-(0-0.6)}Br_{(0-0.6)}$ was developed[108] (μτ product = 0.12 cm^2/V). Crystalline non-lead halide–$Cs_2AgBiBr_6$ demonstrates strong potential toward radiation detection, with XRD confirming cubic Fm3 symmetry and low resistivity (10^6–10^9Ω cm). The material also exhibits high charge-carrier mobilities (upto 4.82 $cm^2/V/s$), meeting the requirements for effective radiation sensors.[109]

7.4 CONCLUSION

The chapter concludes that through their adaptable characteristics and capacity to modify their structure for many kinds of uses, perovskite materials have evolved into revolutionary elements in the discipline of optoelectronics. In devices (such as, photovoltaic cells, sensors, electroluminescent diodes, photodetectors, and lasers, these materials perform outstandingly attributable to their amazing carrier mobilities, tunable bandgaps, and high power conversion efficiencies. Their potential for high-performance and eco-friendly optoelectronic applications is highlighted by their simple production techniques and the promise of lead-free and eco-friendly substitutes. Notwithstanding issues with stability, toxicity, and consistency of performance, developments in perovskite engineering, such as compositional changes, heterostructure designs, and defect passivation, suggest that these materials will have a bright future in commercial applications.

REFERENCES

1. Li, B., Tian, F., Cui, X., Xiang, B., Zhao, H., Zhang, H., Wang, D., Li, J., Wang, X., Fang, X., Qiu, M., & Wang, D., 2022. Review for

rare-earth-modified perovskite materials and optoelectronic applications. *Nanomaterials*, 12(10), 1773.

2. Belich, N., Udalova, N., Semenova, A., Petrov, A., Fateev, S., Tarasov, A., & Goodilin, E., 2020. Perovskite puzzle for revolutionary functional materials. *Front. Chem.*, 8, 550625.
3. Szuromi, P., & Grocholski, B., 2017. Natural and engineered perovskites. *Science*, 358(6364), 732–733.
4. Raveau, B., 2007. The perovskite history: More than 60 years of research from the discovery of ferroelectricity to colossal magnetoresistance via high TC superconductivity. *Prog. Solid State Chem.*, 35(2–4), 171–173.
5. Shcheka, S. S., & Keppler, H., 2012. The origin of the terrestrial noble-gas signature. *Nature*, 490(7421), 531–534.
6. Sarkar, C., Storey, C.D., & Hawkesworth, C.J., 2013. *Detailed Protracted Crystallization History of Perovskite in Orapa Kimberlite. Proc. Int. Kimberlite Conf.*, 211–224. 10.1007/978-81-322-1170-9_13.
7. Chen, W., & Simonetti, A., 2014. Evidence for the Multi-Stage Petrogenetic History of the Oka Carbonatite Complex (Québec, Canada) as Recorded by Perovskite and Apatite. *Minerals*,4(2), 437–476.
8. Park, N.-G., 2018. Halide perovskite photovoltaics: History, progress, and perspectives. *MRS Bull.*, 43(7), 527–533.
9. Wojciechowski, K., Stranks, S. D., Abate, A., Sadoughi, G., Sadhanala, A., Kopidakis, N., Rumbles, G., Li, C.-Z., Friend, R. H., Jen, A. K.-Y., & Snaith, H. J., 2014. Heterojunction modification for highly efficient organic–inorganic perovskite solar cells. *ACS Nano, 8*(12), 12701–12709.
10. Li, G. P., Lu, S. Q., Chen, X., Liao, W. Q., Tang, Y. Y., & Xiong, R. G., 2019. A Three-Dimensional M_3AB-Type Hybrid Organic–Inorganic Antiperovskite Ferroelectric: $[C_3H_7FN]_3$ $[SnCl_6]$ Cl. *Chem. Eur. J.*, *25*(72), 16625-16629.
11. Prasatkhetragarn, A., Sareein, T., Triamnak, N., & Yimnirun, R., 2022. Dielectric and ferroelectric properties of modified-$BaTiO_3$ lead-free ceramics prepared by solid solution method. *Ferroelectrics*, *586*(1), 224–241.
12. Bhalla, A.S., Guo, R., & Roy, R., 2000. The perovskite structure—a review of its role in ceramic science and technology. *Mater. Res. Innov.*, 4(1), 3–26.
13. Cao, R.-Z., Zhang, L.-J., Ding, L.-Y., Liu, X.-P., Liu, X.-Y., Jin, P., Liu, S.-T., & Tao, H.-C., 2022. Transformation from a high-temperature superconductor yttrium-barium-copper oxide to a photocatalytic material yttrium-barium oxide, a DFT-based investigation, *Comput. Mater. Sci.*, 213, 111558.
14. Green, M. A., & Ho-Baillie, A., 2017. Perovskite solar cells: The birth of a new era in photovoltaics. *ACS Energy Letters, 2*(4), 822–830.
15. Wang, P., Guo, Y., Yuan, S., Yan, C., Lin, J., Liu, Z., & Cai, C., 2016. Advances in the structure and materials of perovskite solar cells. *Res. Chem. Intermed.*, 42, 625–639.
16. Dai, X., Xu, K., & Wei, F., 2020. Recent progress in perovskite solar cells: the perovskite layer. *Beilstein J. Nanotechnol.*,11, 51–60.

17. Kim, J.Y., Lee, J.-W., Jung, H.S., Shin, H., &Park, N.-G., 2020. High-efficiency perovskite solar cells. *Chem. Rev.*, 120(15), 7867–7918.
18. Gjorgjevikj, K., Bukleski, M., Dimitrovska-Lazova, S., & Aleksovska, S., 2023. Raman spectroscopy as a tool for predicting the dimensionality and octahedra connectivity in lead iodide perovskites. *J. Mol. Struct.*,1293, 136236.
19. Sharma, R., Menahem, M., Dai, Z., Gao, L., Brenner, T.M., Yadgarov, L., Zhang, J., Rakita, Y., Korobko, R., Pinkas, I., Rappe, A.M., & Yaffe, O., 2020. Lattice mode symmetry analysis of the orthorhombic phase of methylammonium lead iodide using polarized Raman. *Phys. Rev. Mater.*, 4(5), 051601.
20. Satta, J., Melis, C., Carbonaro, C.M., Pinna, A., Salado, M., Salazar, D., & Ricci, P.C., 2021. Raman spectra and vibrational analysis of CsPbI3: A fast and reliable technique to identify lead halide perovskite polymorphs. *J. Mater.*, 7(1), 127–135.
21. Salado, M., Salazar, D., Satta, J., Pinna, A., Melis, C., Carbonaro, C.M., & Ricci, P.C., 2021. Characterization of the ammonium bending vibrations in two-dimensional hybrid lead-halide perovskites from Raman spectroscopy and first-principles calculations. *J. Phys. Chem. C.*, 125(2), 85–92.
22. Chen, X., Xu, Y., Wang, Z., Wu, R. J., Cheng, H. L., & Chui, H. C., 2022. Characterization of a $CH_3NH_3PbI_3$ perovskite microwire by Raman spectroscopy. *J. Raman Spectrosc.*, *53*(2), 288–296.
23. Huong, D.T.M., Nam, N.H., Vu, L.V., & Long, N.N., 2012. Preparation and optical characterization of Eu^{3+}-doped $CaTiO_3$ perovskite powders. *J. Alloys Compd.*, 537, 54–59.
24. Cascos, V., Martínez-Coronado, R., Fernández-Díaz, M. T., & Alonso, J. A., 2020. Topotactic oxidation of perovskites to Novel $SrMo_{1-x}M_xO_{4-\delta}$ (M = Fe and Cr) deficient Scheelite-type oxides. *Materials*, *13*(19), 4441.
25. Kojima, A., Teshima, K., Shirai, Y., & Miyasaka, T., 2009. Organometal halide perovskites as visible-light sensitizers for photovoltaic cells. *J. Am. Chem. Soc.*, 131(17), 6050–6051.
26. Zhang, H., Ji, X., Yao, H., Fan, Q., Yu, B., & Li, J., 2022. Review on efficiency improvement effort of perovskite solar cell . *Sol. Energy*, 233, 421–434.
27. Pandey, S., Shukla, A., & Tripathi, A., 2023. Improvement in the performance and efficiency on self-deficient $CaTiO_3$: Towards sustainable and affordable new-generation solar cells. *Mater. Today Commun.*, *35*, 105575.
28. Duan, L., Zhang, H., Liu, M., Grätzel, M., & Luo, J., 2022. Phase-Pure γ-$CsPbI_3$ for efficient inorganic perovskite solar cells. *ACS Energy Lett.*, 7(9), 2911-2918.
29. Grischek, M., Caprioglio, P., Zhang, J., Pena-Camargo, F., Sveinbjörnsson, K., Zu, F., Menzel, D., Warby, J.H., Li, J., Koch, N. & Unger, E., 2022. Efficiency potential and voltage loss of inorganic $CsPbI_2Br$ perovskite solar cells. *Sol. RRL.*, *6*(11), 2200690.
30. Kim, G., Kwon, N., Lee, D., Kim, M., Kim, M., Lee, Y., Kim, W., Hyeon, D., Kim, B., Jeong, M.S., Hong, J & Yang, J., 2022. Methylammonium

compensation effects in $MAPbI_3$ perovskite solar cells for high-quality inorganic CuSCN hole transport layers. *ACS Appl. Mater. Interfaces*, *14*(4), 5203-5210.

31. Ahmad, K., Kumar, P., Kim, H., & Mobin, S. M., 2022. Optoelectronic and Photovoltaic Properties of $(NH_4)_3Bi_2I_9$: A Perovskite-like Energy Material for Pb-free Perovskite Solar Cells. *ChemNanoMat*, *8*(6), e202200061.
32. Huang, J., Wang, H., Jia, C., Yang, H., Tang, Y., Gou, K., Zhou, Y & Zhang, D., 2024. High-Efficiency and Ultra-Stable Cesium–Bismuth-Based Lead-free Perovskite Solar Cells without Modification. *J. Phys. Chem. Lett.*, *15*(12), 3383–3389.
33. Sabbah, H., Arayro, J., *&* Mezher, R., 2022. Numerical simulation and optimization of highly stable and efficient lead-free perovskite $FA_{1-x}Cs_xSnI_3$-based solar cells using SCAPS. *Materials*, *15*(14), 4761.
34. Hossain, M. K., Samajdar, D. P., Das, R. C., Arnab, A. A., Rahman, M. F., Rubel, M. H. K., Islam, M.R., Bencherif, H., Pandey, R., Madan, J. & Mohammed, M. K., 2023. Design and simulation of Cs_2BiAgI_6 double perovskite solar cells with different electron transport layers for efficiency enhancement. *Energy & Fuels*, *37*(5), 3957–3979.
35. Meng, G., Elumalai, N. K., Mehdizadeh-Rad, H., Ram, K. S., Setsoafia, D. D. Y., & Ompong, D., 2024. Investigating the impact of interfacial layers on device performance of highly stable $Cs_2InBiBr_6$ based double perovskite solar cells. *Adv. Theory Simul.*, 7(2), 2300784.
36. Qin, Z., Pols, M., Qin, M., Zhang, J., Yan, H., Tao, S., & Lu, X., 2023. Over-18%-efficiency quasi-2D ruddlesden–popper Pb–Sn mixed perovskite solar cells by compositional engineering. *ACS Energy Lett., 8*(7), 3188–3195.
37. Eremeev, K., Loiko, P., Zhao, C., Lin, Z. L., Mateos, X., Zin Elabedine, G., Camy, P., Braud, A., Griebner, U., Petrov, V. & Chen, W., 2024. Growth, spectroscopy and laser operation of Tm, Ho: $GdScO_3$ perovskite crystal. *Opt. Express*, *32*(8), 13527–13542.
38. Guo, P., An, M., Shu, Y., Peng, X., Han, Y., Hu, H., Chen, W., Zhao, Y., Li, X. & Li, J., 2022. Nonlinear photonics device based on double perovskite oxide Ba_2LaTaO_6 for ultrafast laser generation. *Opt. Laser Technol., 155*, 108334.
39. He, Y., Su, Z., Cao, F., Cao, Z., Liu, Y., Zhao, C., Weng, G., Hu, X., Tao, J., Chu, J., & Akiyama, H., 2023. Lasing properties and carrier dynamics of $CsPbBr_3$ perovskite nanocrystal vertical-cavity surface-emitting laser. *Nanophotonics*, *12*(12), 2133–2143.
40. Laskar, S., & Sudakar, C., 2024. Photostable $Cs_{0.5}Rb_{0.5}PbBr_3$ Quantum Dots for Whispering Gallery Mode Lasing. *ACS Appl. Opt. Mater.*, 2(8), 1591–1599.
41. Zhan, Z., Liu, Z., Du, J., Huang, S., Li, Q., Hu, Z., Luo, J., Yang, Y., Dong, S., Wang, L. & Tang, J., 2023. Thermally evaporated $MAPbBr_3$ perovskite random laser with improved speckle-free laser imaging. *ACS photonics*, *10*(9), 3077–3086.
42. Si, S., Guo, X., Gan, W., Zhang, X., Lan, B., Li, M. R., & Wang, J., 2022. Lead-free perovskite variant Rb_2SnCl_6: Te based phosphor-sapphire composite for high-power laser-driven lighting. *J. Lumin.*, *251*, 119212.

43. Chen, Y., Li, Z., Tang, T., Cheng, Y., Cheng, L., Wang, X., Haidry, A.A., Jannat, A., & Ou, J. Z., 2024. Room-temperature optoelectronic NO_2 sensing using two-dimensional gallium oxyselenides. *ACS Appl. Nano Mater.*, *7*(3), 3229–3238.
44. Li, X., Liu, W., Gao, Y., Qin, Y., Long, H., Wang, K., Wang, B., & Lu, P., 2022. Two-photon-pumped amplified spontaneous emission from Ruddlesden-Popper perovskite flakes. *Opt. Express*, *30*(12), 21094–21102.
45. Zhang, Y., Dong, Z., Guo, X., Hu, Y., Zhang, Z., Deng, Y., Zhang, Y., Men, Z., Geng, C., Wang, Y., & Chen, Z., 2024. Enhanced performance of amplified spontaneous emission in Dion-Jacobson phase quasi-2D perovskite by facilitating carrier co-radiation. *Opt. Express*, *32*(15), 26306–26317.
46. Du, P., Li, J., Wang, L., Liu, J., Li, S., Liu, N., Li, Y., Zhang, M., Gao, L., Ma, Y., & Tang, J., 2019. Vacuum-deposited blue inorganic perovskite light-emitting diodes. *ACS Appl. Mater. Interfaces*,11, 50, 47083–47090.
47. Hoyeon, K., Roh, K., Murphy, J.P., Zhao, L., Gunnarsson, W.B., Longhi, E., Barlow, S., Marder, S., Rand, B.P., & Giebink, N., 2019. Optically pumped lasing from hybrid perovskite light-emitting diodes. *Adv. Opt. Mater.*, 8(1),1901297.
48. Gunnarsson, W.B., & Rand, B.P., 2020. Electrically driven lasing in metal halide perovskites: Challenges and outlook . *APL Mater.*, 8, 030902.
49. Wu, T., Li, J., Zou, Y., Xu, H., Wen, K., Wan, S., Bai, S., Song, T., McLeod, J.A., Duhm, S., Gao, F., & Sun, B., 2020. High-performance perovskite light-emitting diode with enhanced operational stability using lithium halide passivation. *Angew. Chem. Int. Ed. Engl.*, *59*(10), 4099–4105.
50. Guan, Z., Li, Y., Zhu, Z., Zeng, Z., Shen, D., Tan, J., Tsang, S.-W., Liu, S., & Lee, C.-S., 2021. Efficient perovskite white light-emitting diode based on an interfacial charge-confinement structure. *ACS Appl. Mater. Interfaces*,13(37), 44991–45000.
51. Wang, K., Jin, L., Gao, Y., Liang, A., Finkenauer, B.P., Zhao, W., Wei, Z., Zhu, C., Guo, T.-F., Huang, L. & Dou, L., 2021. Lead-free organic–perovskite hybrid quantum wells for highly stable light-emitting diodes. *ACS Nano*, 15(4), 6316–6325.
52. Tang, H., Xu, Y., Hu, X., Hu, Q., Chen, T., Jiang, W., Wang, L. & Jiang, W., 2021. Lead-free halide double perovskitenanocrystals for light-emitting applications: strategies for boosting efficiency and stability. *Adv. Sci.*,8, 2004118.
53. Kim, H., Bae, S.-R., Lee, T., Lee, H., Kang, H., Park, S., Jang, H.W., & Kim, S.Y., 2021. Enhanced optical properties and stability of $CsPbBr_3$ nanocrystals through nickel doping. *Adv. Funct. Mater.*, 31(28), 2102770.
54. Chen, Z., Li, Z., Chen, Z., Xia, R., Zou, G., Chu, L., Su, S.J., Peng, J., Yip, H.L., & Cao, Y., 2021. Utilization of trapped optical modes for white perovskite light-emitting diodes with efficiency over 12%. *Joule*, 5, 456–466.
55. Yang, J.N., Chen, T., Ge, J., Wang, J.J., Yin, Y.C., Lan, Y.F., Ru, X.C., Ma, Z.Y., Zhang, Q., & Yao, H.B., 2021. High color purity and efficient green light-emitting diode using perovskitenano crystals with the size overly exceeding bohrexciton diameter. *J. Am. Chem. Soc.*, 143(47), 19928–19937.

56. Wang, F., Zou, X., Xu, M., Wang, H., Wang, H., Guo, H., Guo, J., Wang, P., Peng, M., Wang, Z., & Wang, Y., 2021. Recent progress on electrical and optical manipulations of perovskite photodetectors. *Adv. Sci.*,8(14), e2100569.
57. Nazari, N. F., Rajabi, M., & Moshfegh, A. Z., 2024. UV-activated heterojunction in $BaTiO_3$ decorated ZnO nanorods for faster and more efficient photodetector. *Chem. Eng. J.*, 379, 115877.
58. Abbas, Z., Parveen, A., & Al-Enizi, A.M., 2024. Investigating the effect of Ca-doping concentration on optical and electronic properties of $RaTiO_3$ for energy applications. *Chem. Pap.*, 78(16), 8821–8832.
59. Zhang, Y., Yao, J., Zhang, Z., Zhang, R., Li, L., Teng, Y., Shen, Z., Kang, L., Wu, L., & Fang, X., 2023. Two-dimensional perovskite $Pb_2Nb_3O_{10}$ photodetectors. *J. Mater. Sci. Technol.*, 164, 95–101.
60. Liu, X., Li, S., Li, Z., Zhang, Y., Yang, W., Li, Z., Liu, H., Shtansky, D.V. & Fang, X., 2021. Boosted responsivity and tunable spectral response in B-site substituted 2D $Ca_2Nb_{3-x}Ta_xO_{10}$ perovskite photodetectors. *Adv. Funct. Mater.*, 31(20), 2101480.
61. Liu, X., Li, S., Li, Z., Cao, F., Su, L., Shtansky, D. V., & Fang, X., 2022. Enhanced response speed in 2D perovskite oxides-based photodetectors for UV imaging through surface/interface carrier-transport modulation. *ACS Appl. Mater. Interfaces*, 14(43), 48936–48947.
62. Tian, W., Zhou, H., & Li, L., 2017. Hybrid organic–inorganic perovskite photodetectors. *Small*, 13(41), 1701988.
63. Liu, L., Xu, M., Cheng, P., Tian, X., Xu, X., Tao, X., & Gao, Z., 2024. Isotope synergetic effect on hybrid organic–inorganic perovskite single crystal carrier transport and high-performance photodetectors. *ACS Mater. Lett.*,6(8), 3834–3843.
64. Huang, L., Wang, Y., Zhu, X., Zhao, X., Li, G., Li, L., & Sun, W., 2021. Mg-doped nickel oxide as efficient hole-transport layer for perovskite photodetectors. *Phys. Chem. Chem. Phys.*,125(29), 16066–16074.
65. Zhou, H., Ren, D., Huang, J., Xiao, B., & Wang, H., 2022. Monolayer cesium lead bromine chloride nanocrystal self-powered photodetectors. *IEEE Electron Device Lett.*, 43(8),1283–1286.
66. Wang, Y., Bai, S., Liang, H., Li, C., Sun, J., Li, X., Tan, T., Yang, G., & Wang, J., 2023. Cesium tin halide perovskite quantum dots for high-performance ultraviolet photodetectors. *J. Lumin.*, 257, 119700.
67. Shuang, Z., Zhou, H., Wu, D., Zhang, X., Xiao, B., Duan, J., & Wang, H., 2022. High-performance Ag_2BiI_5 Pb-free perovskite photodetector. *Photonics Res.*, 10(8), 1886–1891.
68. Dai, Z.Y., Chen, C., Wang, G.S., Lyu, Y.N., & Ma, N., 2023. Bandgap-tuned barium bismuth niobate double perovskite for self-powered photodetectors with a full-spectrum response. *J. Mater. Chem. C.*, 11, 574–582.
69. Ding, N., & Shao, L., 2022. Highly-sensitive, stable, and fast-response lead-free $Cs_2AgBiBr_6$ double perovskite photodetectors enabled by synergistic engineering of doping Na^+/Ce^{3+} and integrating Ag nanoparticles film. *Laser Photonics Rev.*, 16(12), 2200291.

70. Xue, W., Pan, G., Li, C., Zhu, Y., Wang, Y., Kong, B., Gao, H., Zhang, H., You, W., & Mao, Y., 2024. Thulium-doped sodium-ytterbium double perovskite single crystals: A novel and efficient upconverter for achieving highly sensitive narrowb and near-infrared perovskite photodetectors. *Adv. Opt. Mater.*, 12(6), 2301873.
71. Fan, Q., Ma, Y., Xu, H., Liu, Y., Wang, B., Guo, W., Hu, X., Luo, J., & Sun, Z., 2023. Acquiring a newly tailored 2D Dion–Jacobson hybrid perovskite with large structural distortion for efficient crystal array photodetector. *Adv. Opt. Mater.*, 11(2), 2202277.
72. Kokulnathan, T., & Wang, T.J., 2019. Synthesis and characterization of 3D flower-like nickel oxide entrapped on boron doped carbon nitride nanocomposite: An efficient catalyst for the electrochemical detection of nitrofurantoin. *Compos. B Eng.*, 176, 106914.
73. Kokulnathan, T., Suvina, V., Wang, T., & Balakrishna, R.G., 2019. Synergistic design of a tin phosphate-entrapped graphene flake nanocomposite as an efficient catalyst for electrochemical determination of the antituberculosis drug isoniazid in biological samples. *Inorg. Chem. Front.*, 6(7), 1831–1841.
74. Guziejewski, D. 2019. Square-wave amplitude effect in cathodic and anodic stripping square-wave voltammetry. *Electroanalysis*, 31(2), 231–238.
75. Muthukutty, B., Karthik, R., Chen, S., & Abinaya, M., 2019. Designing novel perovskite-type strontium stannate ($SrSnO_3$) and its potential as an electrode material for the enhanced sensing of anti-inflammatory drug mesalamine in biological samples. *New J. Chem.*, 43, 12264–12274.
76. Muthumariyappan, A., Rajaji, U., Chen, S.M., Baskaran, N., Chen, T.W., & JothiRamalingam, R., 2019. Sonochemical synthesis of perovskite-type barium titanate nanoparticles decorated on reduced graphene oxide nanosheets as an effective electrode material for the rapid determination of ractopamine in meat samples. *Ultrason.Sonochem.*,56, 318–326.
77. Anajafi, Z., Naseri, M., Marini, S., Espro, C., Iannazzo, D., Leonardi, S.G., & Neri, G., 2019. $NdFeO_3$ as a new electrocatalytic material for the electrochemical monitoring of dopamine. *Anal.Bioanal. Chem.*, 411, 7681–7688.
78. Akbari, S., Foroughi, M.M., Nadiki, H.H., & Jahani, S., 2019. Synthesis and characterization of $LaMnO_3$ nanocrystals and graphene oxide: fabrication of graphene oxide–$LaMnO_3$ sensor for simultaneous electrochemical determination of hydroquinone and catechol. *J. Electrochem. Sci. Eng.*, 9, 255–267.
79. Atta, N.F., El-Ads, E.H., Galal, A., & Galal, A.E., 2019. Electrochemical sensing platform based on nano-perovskite/glycine/carbon composite for amlodipine and ascorbic acid drugs. *Electroanalysis*, 31(3), 448–460.
80. Suvina, V., Kokulnathan, T., Wang, T.J., & Balakrishna, R.G., 2020. Unraveling the electrochemical properties of lanthanum cobaltite decorated halloysite nanotube nanocomposite: An advanced electrocatalyst for determination of flutamide in environmental samples. *Ecotoxicol. Environ. Saf.*, 190, 110098.
81. He, J., Sunarso, J., Miao, J., Sun, H., Dai, J., Zhang, C., Zhou, W., & Shao, Z., 2019. A highly sensitive perovskite oxide sensor for detection of p-phenylenediamine in hair dyes. *J. Hazard. Mater.*, 369, 699–706.

82. Ahmad, K., Kumar, P., & Mobin, S.M., 2020. A highly sensitive and selective hydroquinone sensor based on a newly designed N-rGO/$SrZrO_3$ composite. *Nanoscale Adv.*, 2(1), 502–511.
83. Sundaresan, P., Karthik, R., Chen, S.M., Vinoth Kumar, J., Muthuraj, V., & Nagarajan, E.R., 2019. Ultrasonication-assisted synthesis of sphere-like strontium cerate nanoparticles ($SrCeO_3$ NPs) for the selective electrochemical detection of calcium channel antagonists nifedipine. *Ultrason. Sonochem.*, 53, 44–54.
84. Cao, F., Dong, Q., Li, C., Chen, J., Ma, X., Huang, Y., Song, D., Ji, C., & Lei, Y., 2018. Electrochemical sensor for detecting pain reliever/fever reducer drug acetaminophen based on electrospun $CeBiO_x$ nanofibers modified screen-printed electrode. *Sens. Actuators B Chem.*, 256, 143–150.
85. Li, X., Dar, M.I., Yi, C., Luo, J., Tschumi, M., Zakeeruddin, S.M., Nazeeruddin, M.K., Han, H., & Grätzel, M., 2015. Improved performance and stability of perovskite solar cells by crystal crosslinking with alkylphosphonic acid ω-ammonium chlorides. *Nat. Chem.*, 7(9), 703–711.
86. Shen, D., Pang, A., Li, Y., Dou, J., & Wei, M., 2018. Metal–organic frameworks at interfaces of hybrid perovskite solar cells for enhanced photovoltaic properties. *Chem. Commun.*, 54(10), 1253–1256.
87. Zhao, Y., Wei, J., Li, H., Yan, Y., Zhou, W., Yu, D., & Zhao, Q., 2016. A polymer scaffold for self-healing perovskite solar cells. *Nat. Commun.*, 7, 10228.
88. Zuo, L., Guo, H., deQuilettes, D.W., Jariwala, S., De Marco, N., Dong, S., DeBlock, R., Ginger, D.S., Dunn, B., Wang, M., & Yang, Y., 2017. Polymer-modified halide perovskite films for efficient and stable planar heterojunction solar cells. *Sci. Adv.*, 3(8), e1700106.
89. Gomez, L., de Weerd, C., Hueso, J.L., & Gregorkiewicz, T., 2017. Color-stable water-dispersed cesium lead halide perovskite nanocrystals. *Nanoscale*, 9(2), 631–636.
90. Ahn, N., Son, D.-Y., Jang, I.-H., Kang, S.M., Choi, M., & Park, N.G., 2015. Highly reproducible perovskite solar cells with average efficiency of 18.3% and best efficiency of 19.7% fabricated *via* Lewis base adduct of lead (II) iodide. *J. Am. Chem. Soc.*, 137(27), 8696–8699.
91. Xu, C., Liu, Z., & Lee, E., 2019. High-performance inverted planar perovskite solar cells using a pristine fullerene mixture as an electron-transport layer. *J. Mater. Chem. C.*, 7(23), 6956–6963.
92. Ma, C., Lo, M.F., & Lee, C.S., 2018. Stabilization of organometallic halide perovskite nanocrystals in aqueous solutions and their applications in copper ion detection. *Chem. Commun.*, 54(45), 5784–5787.
93. Stoeckel, M.-A., Gobbi, M., Bonacchi, S., Liscio, F., Ferlauto, L., Orgiu, E., & Samorì, P., 2017. Reversible, fast, and wide-range oxygen sensor based on nanostructured organometal halide perovskite. *Adv. Mater.*, 29(38), 1702469.
94. Kakavelakis, G., Gagaoudakis, E., Petridis, K., Petromichelaki, V., Binas, V., Kiriakidis, G., & Kymakis, E., 2017. Solution processed $CH_3NH_3PbI_{3-x}Cl_x$ perovskite based self-powered ozone sensing element operated at room temperature. *ACS Sens.*, 3(1), 135–142.

95. Natkaeo, A., Phokharatkul, D., Hodak, J.H., Wisitsoraat, A., & Hodak, S.K., 2018. Highly selective sub–10 ppm H_2S gas sensors based on Ag-doped $CaCu_3Ti_4O_{12}$ films. *Sens. Actuators B Chem.*, 260, 571–580.
96. Shimizu, Y., Yamamoto, S., & Takase, S., 2017. A thick-film impedancemetric carbon monoxide sensor using layered perovskite-type cuprate. *Sens. Actuators B Chem.*, 249, 667–672.
97. Chakraborty, S., & Pal, M., 2018. Highly efficient novel carbon monoxide gas sensor based on bismuth ferrite nanoparticles for environmental monitoring. *New J. Chem.*, 42(9), 7188–7196.
98. Zhang, X.D., Zhang, W.L., Cai, Z.X., Li, Y.K., Yamauchi, Y., & Guo, X., 2019. $LaFeO_3$ porous hollow micro-spindles for NO_2 sensing. *Ceram.Int.*, 45(5), 5240–5248.
99. Wang, J., Wang, C., Liu, A., You, R., Liu, F., Li, S., Zhao, L., Jin, R., He, J., Yang, Z., Sun, P., Yan, X., & Lu, G., 2019. High-response mixed-potential type planar YSZ-based NO_2 sensor coupled with $CoTiO_3$ sensing electrode. *Sens. Actuators B Chem.*, 287, 185–190.
100. Park, B., Kang, S.M., Lee, G.W., Kwak, C.H., Rethinasabapathy, M., & Huh, Y.S., 2020. Fabrication of $CsPbBr_3$ perovskite quantum dots/cellulose-based colorimetric sensor: dual-responsive on-site detection of chloride and iodide ions. *Ind. Eng. Chem. Res.*, 59(2), 793–801.
101. Xiang, X., Ouyang, H., Li, J., & Fu, Z., 2021. Humidity-sensitive $CsPbBr_3$ perovskite-based photoluminescent sensor for detecting water content in herbal medicines. *Sens. Actuators B Chem.*, 346, 130547.
102. Wang, S., Yu, L., Wei, Z., Xu, Q., Zhou, W., & Xiao, Y., 2022. Dual-response ratiometric fluorescence based ligand-functionalized $CsPbBr_3$ perovskite quantum dots for sensitive detection of trace water in edible oils. *Sens. Actuators B Chem.*, 366, 132010.
103. Zhou, L., Liao, J.-F., Huang, Z.-G., Wei, J.-H., Wang, X.-D., Li, W.-G., Chen, H.-Y., Kuang, D. B., & Su, C.Y., 2019. A highly red-emissive lead-free indium-based perovskite single crystal for sensitive water detection. *Angew. Chem. Int. Ed. Engl.*, 131(16), 5331–5335.
104. Wang, J., Huang, Y., Gao, Z., & Du, J., 2024. Lead-free antimony-based perovskite nanocrystals as a fluorescent nanoprobes for the turn-on detection of trace water in organic solvents. *ACS Appl. Nano Mater.*,7(7), 7958–7965.
105. Ju, Y., Peng, J., Chen, Y., Huang, S., Gu, K., Cai, S., Huang, L., & Zhong, H., 2022. Water-sensitive mixed-phase PEA_6SnI_8 perovskite derivative single crystal for humidity detection. *Cryst. Growth Des.*, 22(8), 4689–4695.
106. Wu, Z., Yang, J., Sun, X., Wu, Y., Wang, L., Meng, G., Kuang, D., Guo, X.Z., Qu, W., Du, B., Liang, C., Fang, X., Tang, X., & He, Y., 2021. An excellent impedance-type humidity sensor based on halide perovskite $CsPbBr_3$ nanoparticles for human respiration monitoring. *Sens. Actuators B Chem.*, 337, 129772.
107. Wei, H., DeSantis, D., Wei, W., Deng, Y., Guo, D., Savenije, T.J., Cao, L., & Huang, J., 2017. Dopant compensation in alloyed $CH_3NH_3PbBr_{3-x}Cl_x$ perovskite single crystals for gamma-ray spectroscopy. *Nat. Mater.*,16, 826–833.

108. Nazarenko, O., Yakunin, S., Morad, V., Cherniukh, I., & Kovalenko, M. V., 2017. Single crystals of caesium formamidinium lead halide perovskites: solution growth and gamma dosimetry. *NPG Asia Mater.*, *9*(4), e373–e373.
109. Murgulov, V., Schweinle, C., Daub, M., Hillebrecht, H., Fiederle, M., Dědič, V., & Franc, J., 2022. Double perovskite $Cs_2AgBiBr_6$ radiation sensor: synthesis and characterization of single crystals. *J. Mater. Sci.* 57, 2758–2774.

Chapter 8

Advanced computational studies of quantum dots for optoelectronic, sensing, and computing applications

Wisdom Benson, Saumika Bandopadhyay, Chase Adams, Basanta Baral, and Prabhakar Misra

8.1 INTRODUCTION

8.1.1 Overview of quantum dots

8.1.1.1 Definition and significance of quantum dots

Quantum dots (QDs) are a class of nanoscale semiconductor particles that have emerged as novel materials in nanotechnology due to their extraordinary optical, electronic, and quantum mechanical properties. These structures typically range in size from a few nanometers to tens of nanometers [1] and are considered zero-dimensional because they confine charge carriers (electrons and holes) in all three spatial dimensions. Such confinement leads to the formation of discrete, quantized energy levels, unlike the continuous energy bands seen in bulk materials. This behavior arises from the "quantum confinement effect" [2], is a direct consequence of the small size of QDs relative to the exciton Bohr radius, which governs the spatial extent of an electron–hole pair in a semiconductor. The size-dependent quantum confinement allows for the precise tuning of key properties, such as emission wavelength and bandgap energy, simply by altering the dimensions of the QD. This unique characteristic has made QDs very useful for nanoscale materials science, offering unprecedented control over material behavior and enabling advancements across diverse applications, including optoelectronics, energy, biology, and quantum information processing.

One of the most notable features of QDs is their ability to display size-tunable optical properties, particularly in their absorption and emission spectra. As the size of a QD decreases, the bandgap energy increases, leading to a blue shift in the emission wavelength. This tunability is particularly valuable in the design of optoelectronic devices, such as light-emitting diodes (LEDs) [3] and lasers, where control over emitted light color is essential. In addition, QDs exhibit high photoluminescence

DOI: 10.1201/9781003512899-8

efficiency, characterized by sharp and intense emission peaks, making them ideal for bioimaging applications, where brightness, stability, and specificity are paramount. Beyond their optical capabilities, QDs exhibit atomic-like behavior with well-defined electronic states and controlled transitions, earning them the nickname [2] "artificial atoms." These properties are being harnessed in cutting-edge quantum technologies, including quantum computing, sensing, and communication, where QDs serve as qubits or single-photon emitters. Moreover, their versatility in synthesis—ranging from elemental QDs like silicon to compound systems like cadmium selenide (CdSe) and zinc oxide (ZnO)—and adaptability in functionalization further enhance their utility in applications such as solar cells, photodetectors, and quantum devices [4].

8.1.1.2 Historical development and advances in nanotechnology

The history of QDs traces back to the early theoretical exploration of quantum confinement effects. In 1982, Efros and Efros postulated that the optical and electronic properties of semiconductors could be controlled by reducing particle size to the nanoscale, resulting in quantized energy levels and tunable emission properties [5]. Shortly thereafter, the synthesis of nanometer-sized semiconductor particles embedded in silicate glasses demonstrated these size-dependent properties, marking one of the earliest experimental confirmations of the quantum confinement effect [6]. The controlled synthesis of cadmium chalcogenide (CdS and CdSe) nanocrystals in the 1980s by Brus and co-workers further advanced the field, allowing for systematic studies of the relationship between size and optical behavior [7].

Throughout the 1990s and 2000s, the development of colloidal synthesis techniques revolutionized QD production by enabling high-quality, monodisperse nanocrystals with tunable properties [8]. Improvements in surface passivation and functionalization enhanced photoluminescence efficiency and stability, facilitating the use of QDs in applications such as bioimaging and LEDs. Concurrently, theoretical models were refined to better describe the electronic structure and quantum mechanical behavior of QDs. In recent decades, research has expanding into novel QD materials, such as perovskites and ZnO, which offer distinct advantages like environmental friendliness and broad bandgap tunability. The integration of QDs into devices for quantum computing, sensing, solar energy harvesting, and single-photon emission underscores their transformative potential, making them an important material for modern nanotechnology applications [9, 10].

8.1.2 Zinc oxide quantum dots

8.1.2.1 Unique properties of ZnO quantum dots

ZnO QDs stand out among semiconductor nanocrystals due to their wide bandgap (~3.37 eV) and high exciton binding energy (~60 meV) [11, 12], which enable efficient excitonic emission at room temperature. These properties make ZnO QDs particularly attractive for ultraviolet (UV) optoelectronic devices, including UV lasers and photodetectors. The quantum confinement effect in ZnO QDs allows for precise tuning of their optical properties, such as photoluminescence and absorption, by varying the particle size. This size dependence shifts the emission wavelength, enabling applications across the UV and visible spectra. In addition, ZnO QDs exhibit a high degree of photostability and low toxicity compared to conventional QDs like CdSe, making them an environmentally friendly alternative for bioimaging and biosensing [1]. Their strong surface polarity and abundance of surface defect states further enhance their sensitivity for sensing applications by enabling selective interactions with target molecules or ions [13].

The high electron mobility of ZnO QDs facilitates their use in electron transport layers in perovskite solar cells and other photovoltaic devices, improving charge collection and overall efficiency. In addition, the wurtzite crystal structure of ZnO contributes to unique spintronic properties, as interactions between acceptor-bound charge carrier spins can potentially mediate room-temperature ferromagnetism, a feature vital for quantum technologies [14]. The nearly spherical geometry of ZnO QDs, with diameters generally less than 10 nm, amplifies optical characteristics by inducing significant quantum confinement effects and altering optical phonon interactions compared to bulk ZnO. These modifications result in highly tunable and sharper photoluminescence, predominantly in the ultraviolet–visible (UV–Vis) spectrum, making ZnO QDs particularly advantageous for biosensing and LED applications. Moreover, the unintentional incorporation of impurity states during the synthesis process often enhances luminescence intensity by introducing defect levels that act as radiative recombination centers [15]. These defect states, coupled with the quantum confinement-induced discrete electronic states, enable precise modulation of optical behavior. Such features allow ZnO QDs to be tailored for specific optoelectronic, photonic, sensing, and spintronic applications.

8.1.2.2 Applications in sensing, optoelectronics, and quantum computing

ZnO QDs have gained significant interest for gas sensing due to their enhanced surface properties, which allow them to interact effectively with various chemical species. The sensing mechanism involves changes in electrical resistance resulting from surface interactions like adsorption and redox processes. When ZnO QDs are reduced to the nanoscale, their surface-to-volume ratio increases, leading to stronger interactions with gaseous molecules, thus improving sensitivity. Research has shown that smaller ZnO QDs (less than 10 nm) exhibit improved performance in gas sensing due to increased adsorption capacity and higher surface reactivity. Advancements in the synthesis of ZnO nanoparticles, including methods like chemical vapor deposition (CVD) and wet chemical synthesis, have further enhanced their gas sensing capabilities [16]. By reducing the particle size, ZnO QDs present a greater density of grain boundaries and interfaces, which significantly improves their reactivity with gaseous species, resulting in more pronounced changes in electrical properties.

In optoelectronic applications, ZnO QDs exhibit superior performance due to their high exciton binding energy (~60 meV), which ensures efficient excitonic emission at room temperature. This, in turn, makes them highly desirable for applications in LEDs and laser diodes, particularly for UV emission. In addition, their tunable optical properties and strong luminescence are enabling the development of next-generation display technologies and photovoltaic devices. ZnO QDs are also used in photodetectors due to their high photosensitivity and rapid response, making them suitable for advanced imaging systems and optical communication technologies.

ZnO QDs are emerging as promising materials in quantum computing as well due to their discrete energy levels, long-lived excitonic states, and potential for spin-based qubit implementation. The high exciton binding energy ensures the stability of excitons at room temperature, critical for quantum coherence in computational applications. ZnO's wurtzite structure [14] and inherent piezoelectricity, combined with the potential for doping with magnetic or rare-earth elements, enable spin manipulation and long spin relaxation times, key requirements for spintronics and quantum information processing.

8.2 FUNDAMENTAL PHYSICS AND CHEMISTRY OF QUANTUM DOTS

In several scientific fields, particularly chemistry, physics, and biology, QDs have become an intriguing material. QDs made up of group II–VI or III–V elements of the periodic table and described as particles having physical dimensions smaller than the exciton's Bohr radius, have captured the imagination of scientists. Since their debut, QDs' utility has grown significantly, and they are now used in a variety of applications, including solar cells, LED technology, lasers, bio-labeling, sensing, and quantum computing.

The exceptional optical and electronic properties of QDs make them highly versatile. These properties include size-tunable photoluminescent (PL) emission, high extinction coefficients, high fluorescent quantum yields, and stability against photobleaching. Furthermore, their small size (4–12 nm in diameter) and fluorescence intermittency (blinking) offer distinct advantages. The size tunability of QDs, in particular, governs their light emission and allows them to be adapted for applications where precision is essential.

8.2.1 Basic quantum dot theory

The features of semiconductor QDs, which have particle sizes in the nanometer range, that are important for optoelectronic applications are not uncommon and need elaboration.

8.2.1.1 The quantum confinement effect

Quantum confinement is the defining feature of QDs. When a semiconductor material is reduced to a few nanometers in size, the movement of charge carriers (electrons and holes) is restricted in all three spatial dimensions. Such spatial restriction forces the energy levels of the material to become discrete (Figure 8.1), akin to those in an atom. Consequently, QDs are often described as "artificial atoms" (this is further discussed in Section 8.3.3).

The degree of confinement dictates the properties of QDs:

- *Strong confinement regime:* In this regime, the QD's radius is significantly smaller than the exciton Bohr radius. The confinement energy—the energy associated with restricting the movement of charge carriers—dominates over the Coulomb interaction between the electron and hole.
- *Weak confinement regime:* When the QD's size exceeds the exciton Bohr radius, Coulomb interactions play a larger role, and confinement effects are less pronounced.

8.2.1.2 Discrete energy levels and their implications

The size of a QD directly affects its bandgap energy. Smaller QDs have larger bandgaps and emit light at shorter wavelengths, while larger QDs

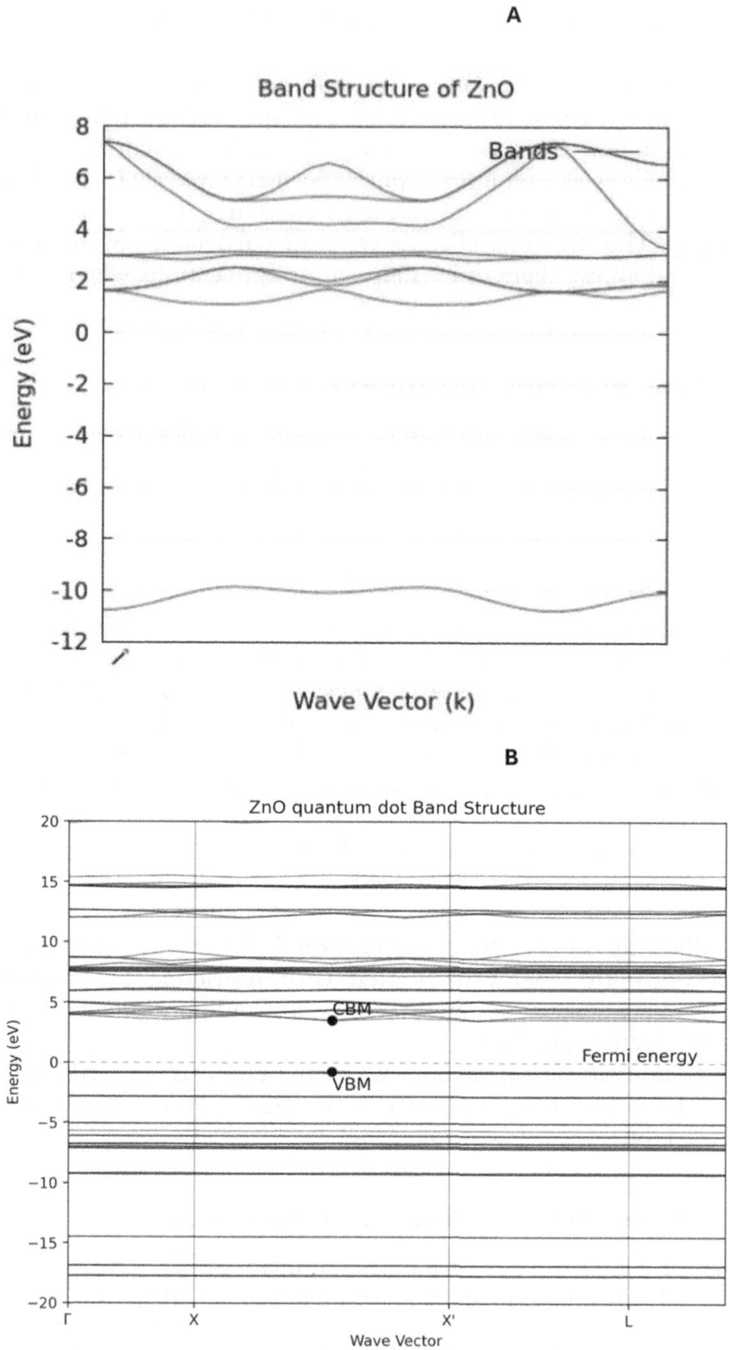
A
Band Structure of ZnO
Bands
8
6
4
2
0
-2
-4
-6
-8
-10
-12
Energy (eV)
Wave Vector (k)
B
ZnO quantum dot Band Structure
20
15
10
5
0
−5
−10
−15
−20
Energy (eV)
CBM
Fermi energy
VBM
Γ
X
X'
L
Wave Vector

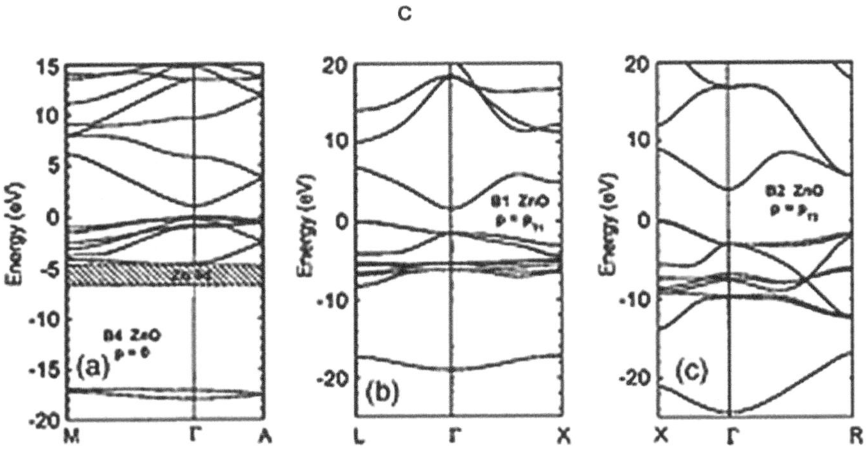

Figure 8.1 A graph with energy bands for bulk materials ZnO (a) transitioning to discrete energy levels in ZnO quantum dots (b). Quantum confinement leads to discrete energy levels in QDs, unlike the continuous bands seen in bulk materials. (c) Band structures for ZnO: a B4 structure at p = 0, b B1 structure at p = pT1, and c B2 structure at p = pT2. (Reprinted with permission from Ref [17]. Copyright 2000 by the American Physical Society.)

have smaller bandgaps and emit at longer wavelengths (Figure 8.2). This property of QDs allows for tuning of their optical properties by increasing or reducing their size, which is critical for applications in optoelectronics and bioimaging.

8.2.1.3 Size-dependent optical properties of qds

This size-bandgap relationship enables optical tunability crucial for device design. Smaller QDs exhibit larger bandgaps, which correspond to higher-energy photon emission (blue-shift). Conversely, larger QDs have smaller bandgaps and emit photons at lower energy (red-shift). Such size-tunable emission is a key feature that distinguishes QDs from other nanomaterials.

1. Advantages of QDs related to size:
   - Comparable size to biological macromolecules, such as nucleic acids and proteins
   - Tunable emission across the visible and near-infrared spectra
   - Enhanced compatibility with biological systems for bio-labeling and imaging

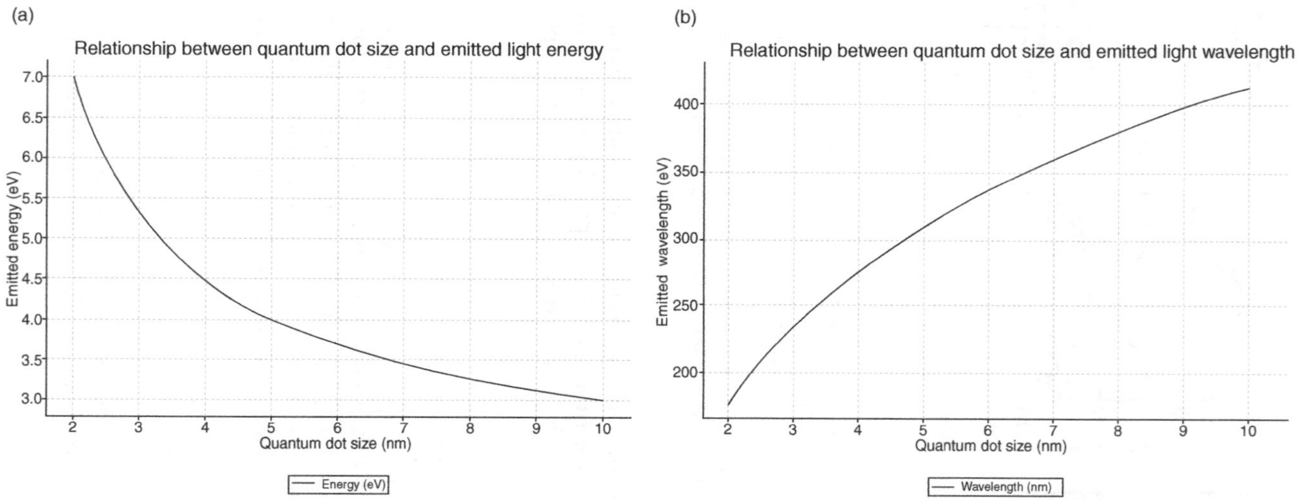

Figure 8.2 The charts above show how the size of the quantum dot and the characteristics of the light it emits are related: (a) Quantum dot size versus emitted energy: As the quantum dot size shrinks, more energy is released. With smaller quantum dots generating higher-energy photons, this inverse connection demonstrates how energy levels change with size. (b) Quantum dot size versus emitted wavelength: The quantum confinement effect is demonstrated by the decreasing wavelength of the emitted light as the quantum dot size decreases. Light with shorter wavelengths (at the blue end of the spectrum) is emitted by smaller quantum dots.

2. *Exceptional features of QDs:* QDs have several advantages that make them highly desirable in scientific and technological applications:
   a. *Small size (4–12 nm in diameter):* Enables integration with biological systems and nanotechnology
   b. *Size-tunable emission:* Facilitates precise wavelength control for applications like LEDs and bio-imaging
   c. *High extinction coefficient:* Enhances light absorption efficiency
   d. *High fluorescent quantum yield:* Ensures bright and stable emission
   e. *Stability against photobleaching:* Allows for long-term imaging and functionality
   f. *Fluorescence intermittency (blinking):* Unique optical behavior that provides insights into single-molecule studies
3. *Theoretical framework:* Several theoretical models describe the behavior of QDs:
   - *Particle in a box model:* Simplifies QDs as particles confined within a potential well, leading to quantized energy states (Figure 8.3)
   - *Effective mass approximation:* Considers the effective masses of electrons and holes in predicting energy levels
   - *Tight-binding models:* Focus on interactions between atomic orbitals to understand the electronic states of QDs

8.2.2 Zinc oxide as a semiconductor

8.2.2.1 Bandgap properties and piezoelectricity

ZnO is a wide bandgap semiconductor with a direct bandgap energy of approximately 3.37 eV at room temperature (~300K) (Figure 8.4), making it ideal for optoelectronic applications, such as UV light emitters and detectors. Its high exciton binding energy (~60 meV) allows for efficient excitonic emissions even at elevated temperatures. Beyond its optical properties, ZnO exhibits remarkable piezoelectricity due to its non-centrosymmetric wurtzite crystal structure. This property enables ZnO to generate an electrical charge when mechanically stressed (Figure 8.4), making it a valuable material for piezoelectric sensors, actuators, and energy harvesting devices.

8.2.2.2 Surface chemistry of zno quantum dots

The surface chemistry of ZnO QDs significantly influences their optical and electronic properties. Due to their high surface-to-volume ratio, ZnO QDs have abundant surface states, including oxygen vacancies, zinc interstitials, and hydroxyl groups [3]. These surface states impact charge carrier dynamics, bandgap energy, and luminescence properties. Surface modifications, such as capping with organic ligands or functionalizing with

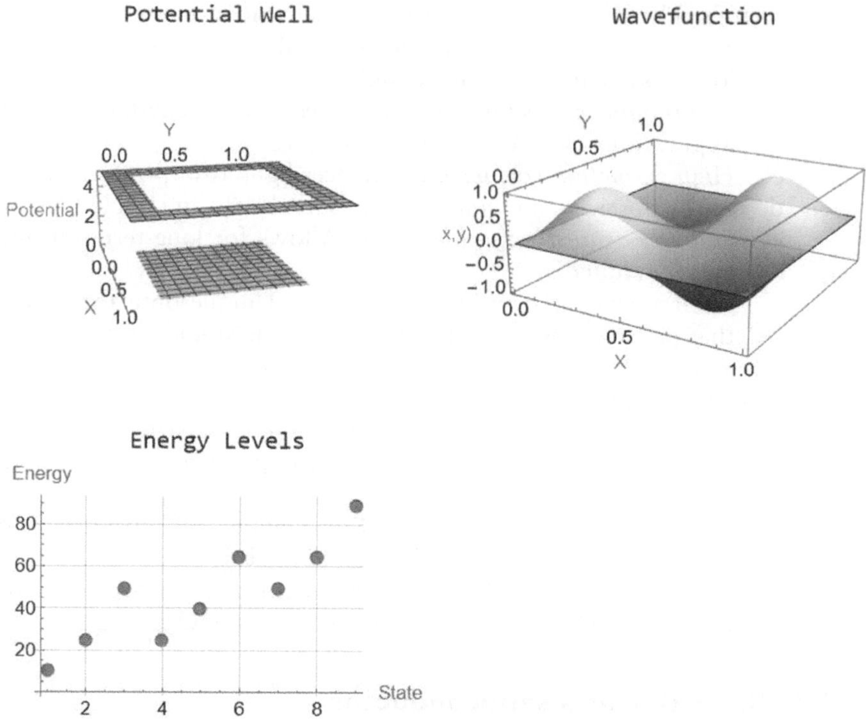

Figure 8.3 Here is a schematic illustration of the "particle in a box" model as applied to a quantum dot. It shows the concept of confinement with discrete energy levels and a wavefunction representation inside the potential barrier. The particle-in-a-box model illustrates the quantization of energy levels in QDs [50].

biomolecules, can tailor the surface properties for specific applications in catalysis, bioimaging, and photovoltaics [4].

Surface defects in ZnO QDs significantly (Figure 8.5) influence their optical and catalytic properties. Oxygen vacancies can introduce localized energy states within the bandgap, affecting photoluminescence and electronic behavior. Surface hydroxylation alters catalytic behavior by modulating adsorption sites, impacting catalytic processes and interactions with other molecules.

8.2.3 Material synthesis and characterization

8.2.3.1 Methods of synthesizing ZnO quantum dots

ZnO QDs can be synthesized through various methods, each offering control over size, shape, and surface properties [5]. Common methods include:

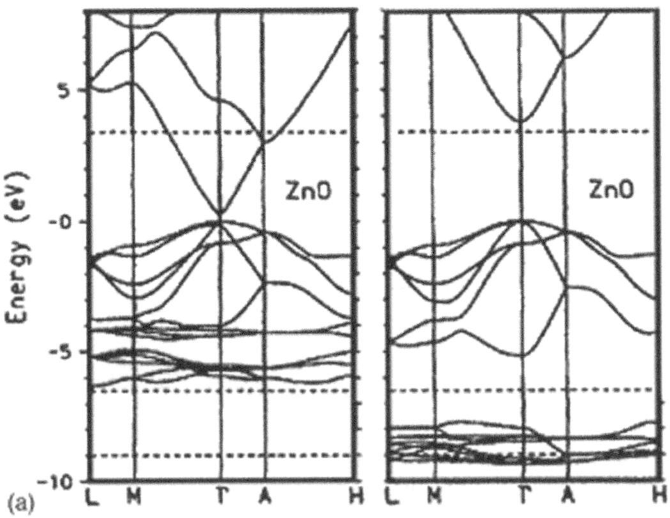

Figure 8.4 (a) Band structure of ZnO showing its wide direct bandgap of 3.37 eV. Caption: "Band structure of ZnO highlighting its 3.37 eV direct bandgap and applications in UV photonics." (b) Schematic of ZnO under mechanical stress illustrating the piezoelectric effect. Piezoelectric effect in ZnO: Mechanical stress induces electrical polarization [53].

Figure 8.5 An SEM image of "comb-like" cantilever arrays of ZnO [54], which are the result of surface polarization induced growth due to the chemically active (0001) Zn. (Image courtesy of Dr XY Kong) [55].

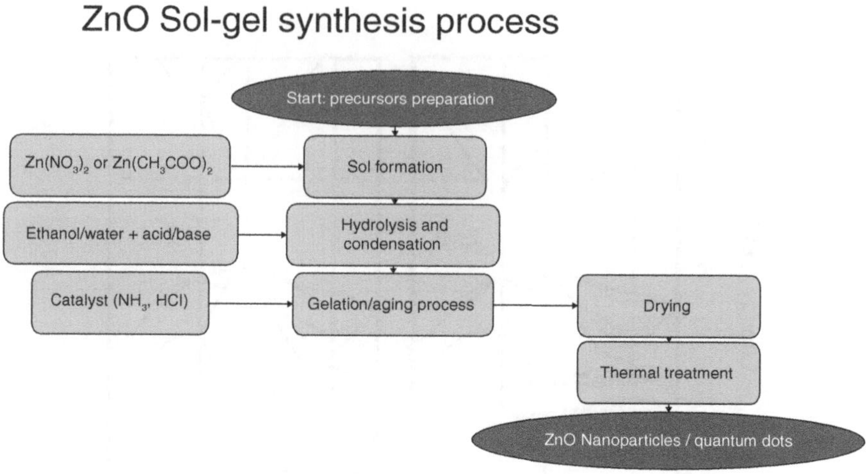

Figure 8.6 A flowchart detailing the sol–gel synthesis process of ZnO quantum dots, showcasing steps such as precursor preparation, sol formation, gelation, drying, and annealing. Steps involved in the sol–gel process for synthesizing ZnO quantum dots [56].

- *Sol–gel process:* A widely used method where zinc precursors undergo hydrolysis and condensation to form ZnO QDs in a colloidal solution (Figure 8.6)
- *Hydrothermal synthesis:* Provides excellent control over crystal size and morphology by utilizing aqueous solutions under high pressure and temperature
- *CVD:* Enables the fabrication of ZnO QDs on substrates for device integration
- *Precipitation method:* A simple, cost-effective approach for producing ZnO QDs in large quantities

8.2.3.2 Surface chemistry of zno quantum dots

The surface chemistry of ZnO QDs can be further engineered post-synthesis to enhance their functionality. Techniques like ligand exchange, salinization, and doping are commonly used to modify their surface. For example, capping ZnO QDs with amines or thiols reduces surface defects and enhances their photoluminescence. In addition, doping with transition metals (e.g., Mn, Fe) introduces magnetic properties, expanding their applications to spintronics and quantum computing [18].

ZnO is a versatile semiconductor with wide-ranging applications due to its unique properties, including a wide bandgap, strong piezoelectricity, and tunable surface chemistry. Advances in synthesis techniques and surface

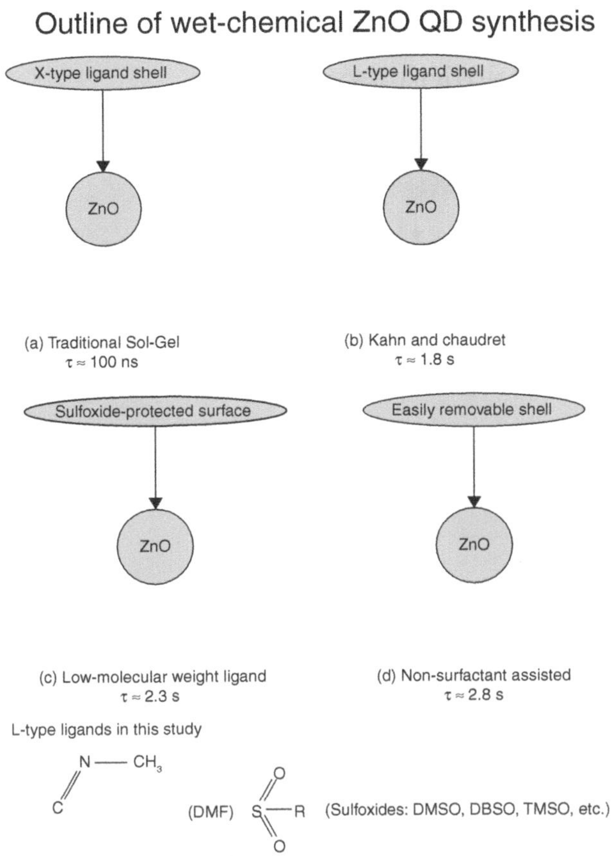

Figure 8.7 An outline of wet-chemical synthetic methods for producing ZnO QDs includes (a) the traditional sol–gel process and organometallic methods, (b) hydrolysis of organometallic precursors using L-type ligands created by Kahn and Chaudret, (c) the OSSOM method, (d) the one-step non-surfactant-assisted method, and (e) low-molecular-weight L-type protector-assisted synthesis [57].

modifications (Figure 8.7) have further expanded its potential in nanotechnology, catalysis, and optoelectronics. Continued research into ZnO QDs promises to unlock new frontiers in science and engineering.

8.3 MODELING AND SIMULATION TECHNIQUES

8.3.1 Density functional theory

Density functional theory (DFT) is a quantum mechanical modeling approach used to investigate the electronic structure of many-body systems

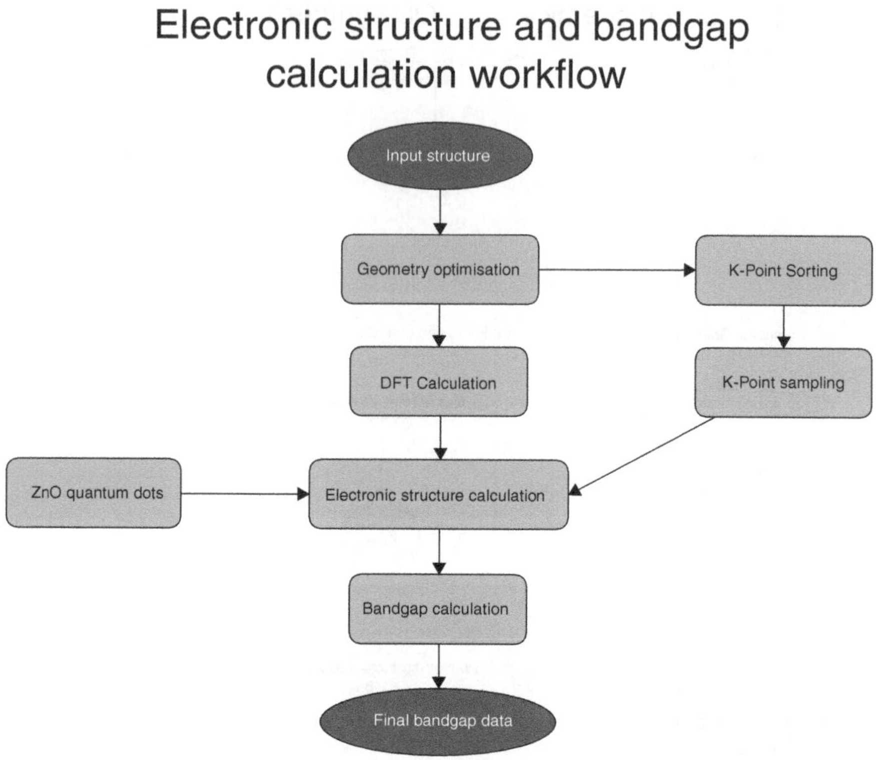

Figure 8.8 Flowchart showing the DFT workflow, from input structure to bandgap calculation, highlighting steps like geometry optimization, k-point sampling, and electronic structure analysis. The DFT workflow for studying ZnO quantum dots [58].

(Figure 8.8). It is widely applied in studying materials like QDs due to its ability to balance computational efficiency and accuracy. At its core, DFT simplifies the many-electron Schrödinger equation by using the electron density $\rho(r)$ as the primary variable instead of the many-electron wavefunction $\Psi(r_1, r_2, \ldots)$.

The total energy of the system is expressed as:

$$E[\rho] = T_s[\rho] + \int v_{ext}(r)\rho(r)dr + J[\rho] + E_{XC}[\rho],$$

where:

- $T_s[\rho]$: Kinetic energy of non-interacting electrons.
- $v_{ext}(r)$: External potential.

- $J[\rho]$: Classical Coulomb energy.
- $E_{XC}[\rho]$: Exchange-correlation energy, incorporating many-body effects.

For ZnO QDs, DFT enables the calculation of bandgaps, density of states (DOS), and surface defect states, critical for predicting their optical and electronic behavior. Modern implementations, such as those using generalized gradient approximations (GGA) or hybrid functionals, provide accurate bandgap predictions and insights into quantum confinement effects. DFT is integral to modeling QDs as it accounts for quantum mechanical interactions efficiently. For example, the bandgap of ZnO, which differs for bulk and QD forms due to confinement, can be accurately predicted by solving the Kohn–Sham equations derived from the above energy functional [19].

8.3.2 Quantum mechanical simulations

Quantum mechanical simulations, including ab initio and semi-empirical methods, delve deeper into the fundamental behavior of electrons in materials. These techniques solve the time-independent Schrödinger equation:

$$\hat{H}\Psi = E\Psi$$

where $\hat{H}$ is the Hamiltonian operator, Ψ the wavefunction, and E the total energy.

In ZnO QDs, simulations focus on:

- *Wavefunction behavior:* Understanding charge carrier dynamics
- *Excitonic properties:* Calculating binding energies of electron-hole pairs
- *Defect states:* Examining the impact of oxygen vacancies and zinc interstitials

Advanced simulations like time-dependent DFT (TD-DFT) extend these calculations to excited-state properties, crucial for optical applications. Tight-binding models and quantum Monte Carlo simulations are also used for larger QDs, balancing precision with computational feasibility. Quantum mechanical simulations are essential for capturing phenomena that emerge at the nanoscale, such as the splitting of energy levels due to quantum confinement and the role of surface states in modifying optical properties [20].

8.3.3 Quantum confinement and band structure

Quantum confinement occurs when the size of a material is reduced to the nanoscale, constraining charge carrier motion. This phenomenon significantly alters the electronic band structure, leading to discrete energy levels instead of continuous bands. For ZnO QDs, the relationship between the size $(d)$of the dot and the bandgap energy E_g can be modeled as:

$$E_g(d) = E_g^{bulk} + \frac{\hbar^2 \pi^2}{2\mu d^2},$$

where:

- E_g^{bulk}: Bandgap of bulk ZnO
- $\hbar$: Reduced Planck's constant
- μ: Effective mass of the electron–hole pair

As (d) decreases, E_g increases, resulting in a blue shift of the emitted light. Such size-dependent bandgap tunability is a hallmark of QDs and has significant implications for optoelectronic devices (Figure 8.9). The effect of quantum confinement on the band structure is critical for applications in photonics and photovoltaics. Smaller ZnO QDs emit UV light due to their widened bandgap and make them suitable for UV LEDs and detectors. Understanding this relationship allows for precise control of QDs over material properties during their synthesis [21].

8.4 PIEZOELECTRICITY IN ZNO QUANTUM DOTS

8.4.1 Link between mechanical stress and electronic response

When some materials are compressed, stretched, or undergo other mechanical deformations, they can produce an electric charge. This phenomenon is known as piezoelectricity. For a material to be piezoelectric, its internal atom arrangement must be irregular or "asymmetrical." This, in turn, allows the electric charges within the material to be separated when it is squeezed, stretched, or otherwise distorted, producing an electrical voltage [22]. We can think of it as follows: water comes out when a sponge is compressed. Similarly, electronic response refers to the redistribution of charge carriers under applied fields or strain, either an electric field or the piezoelectric charge itself ZnO QDs' small size and special characteristics, such as their asymmetrical wurtzite structure, strengthen the link between piezoelectricity and electrical responsiveness. The improved piezoelectric behavior of ZnO QDs is mostly due to the asymmetry of their wurtzite crystal structure

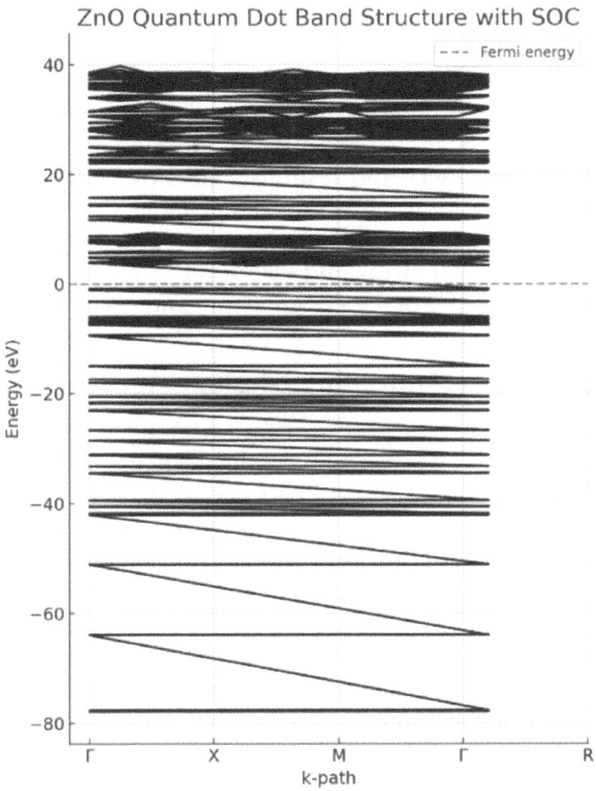

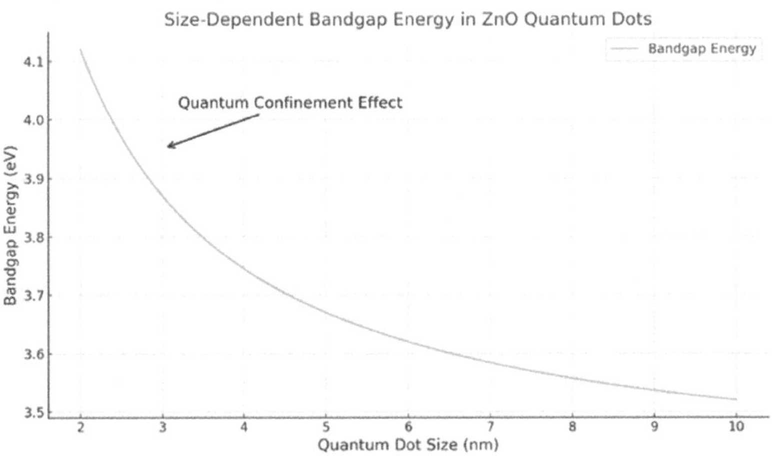

Figure 8.9 The plot of bandgap energy versus quantum dot size for ZnO demonstrates how the bandgap increases as the quantum dot size decreases due to quantum confinement effects.

[23]. The atomic arrangement changes under mechanical stress, producing localized electric fields. These fields then interact with the electrons in the QDs, causing them to move or reorganize. This interaction connects the material's mechanical and electrical characteristics because of its high surface-to-volume ratio; even little mechanical deformations can result in a significant piezoelectric charge, directly impacting electrical response.

This remarkable sensitivity and responsiveness to external forces make ZnO QDs indispensable for applications requiring precision and efficiency, such as advanced sensors and energy-harvesting devices. The unique interaction between mechanical stress and electronic behavior highlights their potential to shape the future of nanoengineering and smart technology [24].

8.4.2 Computational modeling of piezoelectric properties

Computational modeling is a powerful tool for understanding the piezoelectric behavior of ZnO QDs. By using sophisticated methods such as DFT, we can model the effects of mechanical stress on the atomic structure and electrical characteristics of the material. ZnO QDs' atomic interactions and polarization under stress are precisely modeled using exchange-correlation functionals such as local density approximation (LDA) and GGA [25]. The functional choice has significant effects on the results, as LDA tends to underestimate lattice constants, whereas GGA provides a more realistic approximation of ZnO structures. Computing Tool: The following tools from the Quantum Espresso software executables are especially useful when discussing ZnO piezoelectric properties:

1. *ph. x:* Calculates lattice dynamics and phonon modes, which are crucial for piezoelectric response analysis
2. *q2r.x:* Makes it easier to calculate piezoelectric tensors by converting eigenvectors and phonon frequencies into real space
3. *epsilon:* Determines dielectric characteristics, which are closely associated with the piezoelectric effect [26]
4. *matdyn.x:* Connects mechanical stress to electronic behavior by providing the dynamical matrix for phonon dispersion

These simulations are extremely useful for determining the piezoelectric tensor, a mathematical representation of a 3 × 3 matrix that theoretically expresses how mechanical strain in one direction causes electrical polarization in another. For instance, the tensor has the general form:

$$e_{ij} = \begin{bmatrix} e_{11} & e_{21} & e_{31} \\ e_{12} & e_{22} & e_{32} \\ e_{13} & e_{23} & e_{33} \end{bmatrix}$$

where e_{ij} represents the piezoelectric coefficient. It measures how much electric polarization occurs in the *i*-direction due to strain in the *j*-direction [27].

8.4.3 Optoelectronic behavior

Optoelectronic behavior refers to the interaction between light (optical) and electrical properties in a material. In semiconductors like ZnO QDs, this behavior arises from their ability to absorb, emit, and manipulate light, while simultaneously conducting or modulating electrical signals.

8.4.3.1 Bandgap tunability and its implications for device applications

The energy differential between the conduction band, where electrons can flow freely, and the valence band, where they are firmly attached to atoms, is known as the bandgap. This energy gap is not fixed in ZnO QDs but can be modified via the quantum confinement effect, a phenomenon in which electron activity is controlled by the QDs' extremely small size. This tunability results from reducing the size of QDs, which restricts electrons and holes to a smaller volume and modifies their energy levels [28]. As the size of the QDs decreases, the bandgap expands, resulting in higher energy transitions. On the other hand, smaller bandgaps in larger QDs result in lower-energy transitions. Thus, larger QDs emit light in the red or infrared spectrum, whereas smaller ones emit light in the UV or blue spectrum [29]. The exciton Bohr radius of ZnO and the effective mass model control the size-dependent tunability of the bandgap.

The exciton Bohr radius (α_B *) quantifies the distance between an electron and its corresponding hole in the QD. The quantum confinement effects become more prominent when the QD's size gets closer to this radius [30]. The equation for the exciton Bohr radius is given as:

$$\alpha_B * = \epsilon_r \frac{m}{\mu} \alpha_B$$

where α_B is the Bohr radius and m is the mass of the particle,
μ is the reduced mass,
ϵ_r is the size-dependent dielectric constant (relative permittivity).

The effective mass model is given as:

$$m_e^*(confirmed) = m_e^* \left(1 + \frac{\hbar^2}{2m_e^* E_g R^2}\right)$$

where $R = \frac{d}{2} = 5\,\text{nm}$ [31], and E_g is the bulk bandgap of ZnO ($E_g \approx 3.37\,\text{eV}$)

Similarly, for holes:

$$m_h^*(confirmed) = m_h^*\left(1 + \frac{\hbar^2}{2m_h^* E_g R^2}\right)$$

Simplifications:

- $\frac{\hbar^2}{2E_g R^2}$ scales the confinement
- Small confinement results in an increase in m_h^* and m_e^* [32].

8.4.3.2 *Implications for device applications*

1. *Optoelectronic devices:* ZnO QDs' variable bandgap makes them ideal for LEDs and display technologies, allowing fine control over color emission [32].
2. *Solar cells:* By adding ZnO QDs with tunable bandgaps to third-generation solar cells, light absorption is enhanced throughout a wider spectrum, and total efficiency is raised.
3. *Sensing application:* Controlling the bandgap of ZnO QDs improves the sensitivity of chemical and biological sensors, allowing them to detect specific analytes with great precision.
4. *Quantum computing:* QDs can function as qubits, the fundamental units of quantum information, where the quantum states of electrons, such as spin or charge, represent the 0s and 1s of quantum computing.
   The bandgap [33] can be modulated to fine-tune the energy levels inside the QDs, which reduces energy fluctuations and interactions with the environment. This, in turn, is especially crucial for preserving the coherence and stability of qubits.
5. *Energy harvesting:* Devices such as solar systems and piezoelectric nanogenerators perform better in energy-harvesting applications because of the variable bandgaps of ZnO QDs, increasing energy conversion efficiency [32].

8.4.4 Role of applied external fields in modifying optical and electronic responses

ZnO QDs exhibit extremely sensitive optical and electrical characteristics to external fields, including magnetic, electric, and optical fields. These fields have the power to drastically change the interactions between electrons and photons inside the QDs, allowing their characteristics to be tuned for use.

1. *Influence of electric fields:* ZnO QDs can undergo energy level shifts and electron distribution changes when exposed to an electric field. Their optical characteristics, including light emission and absorption, can be altered by this phenomenon, which is the well-known Stark effect [33, 34].

   For example, the applied field can affect the color of light emitted by QDs. It can also amplify or suppress luminescence, making the QDs more sensitive in optoelectronic devices, such as LEDs and lasers.
2. *Role of magnetic fields:* In ZnO QDs, magnetic fields interact with electron spin and orbital mobility to provide precise control over electronic states. This is particularly helpful in quantum computing, where spin states are used as qubits [33]. Furthermore, Zeeman splitting [33, 35], in which energy levels divide into several others, can be induced by magnetic fields. This allows for additional tailoring of the optical and electronic responses and the design of magneto-optical sensors and memory systems.
3. *Optical fields:* ZnO QDs, such as in lasers, interact with optical fields to change their optical sensitivity [27]. Such interaction can either accelerate or suppress optical transitions, which are used in optical switches, photo detectors, and imaging technologies. Because of their high optical field sensitivity, ZnO QDs can be used as precise modulators in sophisticated imaging platforms and communication systems [36]. For example, the absorption and emission properties of the dots can be dynamically adjusted by varying the wavelength or intensity of the input light.

8.5 APPLICATION IN QUANTUM COMPUTING

In this section, we will explore ZnO QDs and their usefulness in quantum computing. But first, what are quantum computers? Quantum computers use "qubits" instead of traditional binary bits that classical computers have. These qubits allow for superposition, meaning instead of either 0 or 1, a bit can hold any floating-point value between and including 0 and 1. Quantum computing is a very prominent and rising field. The advantages quantum computers possess over traditional binary-based computers (superposition of states, quantum entanglement, etc.) allow these devices to solve more complex problems precisely and efficiently [37]. Electronic response refers to the redistribution of charge carriers under applied fields or strain. As ZnO QDs can be controlled through the "Piezoelectric Effect," the ability to generate electrical charge via mechanical stress [34], they are good candidates for holding a multitude of values depending on the amount of stress received.

The outline of this section is as follows: first, we will talk about "nano-sensing" with QDs. Specifically, properties of ZnO QDs for charge-based and spin-based Qubits, and computational modeling of quantum entanglement and coherence. Next, we will discuss data storage and "quantum logic gates," which will cover integration of ZnO QDs into quantum circuits and simulation of charge dynamics and "tunneling behavior."

8.6 NANO-SENSING WITH ZNO QUANTUM DOTS

8.6.1 Properties of ZnO quantum dots for charge-based and spin-based Qubits

ZnO QDs hold many properties that make them good candidates for both charge-based and spin-based qubits. Let us consider charge-based qubits first; ZnO QDs possess what is known as the "quantum confinement effect" [33]. The quantum confinement effect present in these QDs changes their electron energy levels. This creates discrete energy levels, or states, that can represent all the possible values of the qubit (Figure 8.10). This allows control over these states with the use of outer electric fields or "gating" [38] Another phenomenon known as the "Kondo effect" is present in these QDs. This effect emphasizes stronger electron correlation(s), altering both coherence and stability of the states of the qubit. These correlations make

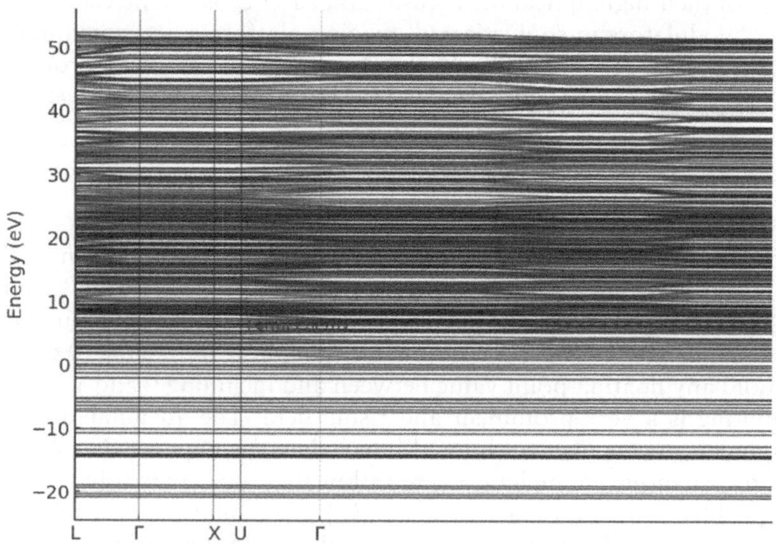

Figure 8.10 The band structure of a ZnO 5 nm (radius) quantum dot. Note the discrete nature of the bands. The ample bandgap gives ZnO properties conducive to quantum computing applications.

the charge-based qubits more robust, especially against background noise/ sound waves coming from the surrounding environment [39].

As far as spin-based Qubits, ZnO, compared to other semiconductors, has a large bandgap (3.37 eV) with a single electron pocket, which prevents "intervalley carrier scattering, weak spin–orbit interaction, and low-density nuclear spins" [40]. These characteristics of ZnO make it ideal for quantum applications that take advantage of long spin coherence. This is perfect, because spin-based qubits use this long spin coherence property to stabilize the system by correcting errors and improve scalability by simplifying spin-based qubit integration in bigger quantum-based systems. Quantum gates also need to maintain coherence to function correctly [41–43].

8.6.2 Computational modeling of quantum entanglement and coherence

ZnO QDs also possess qualities conducive to computational modeling of quantum entanglement, which is a phenomenon present in the study of quantum mechanics. The idea is that particles become so intertwined that a change in the state of one particle influences the state of the other, regardless of the distance between them. This idea challenges prior-held notions about objects needing to remain relatively close to influence one another. Studies have been conducted with the use of computational analysis and simulation to record the emission of photons, coherence times, and probabilities of absorption [44]. Models have also been constructed to perform calculations to find the fidelity of entanglement between ionic systems and ZnO. The results of these studies show ZnO's ability to behave as a mitigator in hybrid quantum systems, using computational modeling to optimize these schemes.

ZnO QDs have a high aptitude for coherence as well. A study that measured spin coherence properties of shallow donor qubits in ZnO found a "Longitudinal Relaxation Time (T_1)" of over 100 ms, indicative of the extraordinary stability of spin states over prolonged periods of time [43]. The same study also measured a "Hahn spin-echo coherence time (T_2)" of about 50 μs. This means ZnO is resistant to decoherence during these quantum functions. These longer coherence times imply that qubits can maintain their states for a longer period, allowing for more complex operations.

8.6.3 Data storage and quantum logic gates

8.6.3.1 Integration of ZnO quantum dots into quantum circuits

ZnO is a very strong candidate for integration into quantum circuits and technologies. QDs may be created within ZnO heterostructures (structures involving a layering of multiple semiconductors) using very accurate voltage control [45]. With this technique, electron states can be manipulated, and phenomena such as the Coulomb diamond and a "parity-independent Kondo effect" are observed [46]. These results indicate that the strong electron correlation within the ZnO structures is important in these phenomena, indicating its potential use in quantum computing and other quantum-based applications.

In fact, significant progress has been made in integrating these dots into quantum systems. Studies have found a "self-assembly" method to arrange ZnO QD molecules in a linear fashion in DNA [47]. These aligned QDs have unique optical properties, which are vital to the development of quantum photonic circuits. The required arrangement of these dots can be achieved through a process called "molecular combing" [48], which is very promising for ZnO QDs being integrated into these circuits for quantum computation applications.

8.6.3.2 Simulation of charge dynamics and tunneling behavior

There has been ample research regarding charge dynamics in ZnO QDs. Studies have explored the electronic states and a phenomenon known as the "Aharonov–Bohm effect" in ZnO QD-ring structures [49]. These dot-ring nanostructures (DRN) are made of a QD that is separated from a concentric quantum ring that circles the dot. This ring is separated from the dot by a "tunneling barrier." DRNs possess properties that have high "tunability," which is important in many applications such as imaging, communications, and environmental monitoring [40, 46]. The DNRs in the study contained few interacting electrons. Unlike comparable DNRs in similar semiconductors, ZnO demonstrates unique traits due to its distinct properties. The energy spectra and Aharonov–Bohm oscillations in ZnO DRNs also depend heavily on the electron number, suggesting that slight changes in size, magnetic field, or potential can change the energy spectra and oscillations greatly [50]. This tunability allows for extensive control over electron charge and spin distribution within the DNR via adjusting the magnetic field or confinement potentials.

ZnO QDs also show evidence of possessing tunneling behavior, which can be classified into three types [45]: *Non-tunneling*, which occurs when the energy distribution lies completely above the potential barrier's maximum, resulting in no tunneling at all; *Full tunneling*, which occurs when the energy distribution of the particle is below the potential barrier minimum, resulting in full tunneling of the particle through the entire barrier; and *Partial tunneling*, which occurs when the particle's energy distribution falls somewhere between the maximum and minimum of the potential barrier. The type of tunneling behavior depends on the nanostructure itself and the surrounding conditions. Very similar to the method used to create QDs within ZnO heterostructures, the Coulomb diamond and the Kondo effect observed from the manipulation of internal states via precise voltage control provides clear evidence of tunneling behavior [38].

8.7 CONCLUSIONS AND FUTURE DIRECTIONS

ZnO QDs have immense potential for a variety of applications, ranging from sensing and optoelectronics to quantum computing, due to their unique optical, electronic, and piezoelectric properties. The ability to tune these properties through size control, doping, and surface modifications has made ZnO QDs an attractive material for next-generation technologies. However, to fully realize their potential, significant challenges remain. Key areas for improvement include enhancing the photoluminescence efficiency and piezoelectric properties of ZnO QDs, which are crucial for their application in advanced optoelectronic devices, sensors, and energy harvesting systems.

The piezoelectricity of ZnO arises from the lack of inversion symmetry in its wurtzite lattice, which enables spontaneous polarization under strain. [14]. These properties are further enhanced when ZnO is reduced to the nanoscale, as quantum confinement effects increase the surface area and grain boundaries, which significantly influence piezoelectric responses. To enhance the piezoelectric properties of ZnO QDs, several strategies are used. One approach is doping with specific elements such as aluminum or gallium, which can increase the material's piezoelectric response by modifying the electronic structure and charge distribution in the lattice [51]. Another strategy is the creation of nanostructured ZnO, such as nanowires [52], which have been shown to exhibit superior piezoelectric performance due to their high aspect ratio and increased surface-to-volume ratio. In addition, improving the alignment of ZnO nanostructures or using composite materials that combine ZnO with other piezoelectric materials could further enhance their overall performance. Improving the stability and reproducibility of ZnO QDs in quantum computing applications is essential. This includes addressing challenges related to error correction, coherence times, and integration with other quantum technologies. Advances in surface passivation techniques [48] and doping strategies can help reduce defect-induced qubit decoherence and improve the performance of ZnO-based quantum computing devices. Research into the development of ZnO QDs with longer coherence times, particularly at room temperature, will be key to scaling these materials for practical quantum information processing applications.

In summary, as properties of materials are further enhanced, ZnO QDs are poised to play a significant role in next-generation technologies, from highly sensitive sensors to scalable quantum computing systems.

ACKNOWLEDGMENTS

Financial support from the National Science Foundation (Award # PHY-1950379, DMR-2101121 & DMR-2349300) and the ACCESS supercomputing allocation support via PHY210066 & DMR190126 and gratefully acknowledged.

REFERENCES

1. Agarwal, K., Rai, H., & Mondal, S. (2023) Quantum dots: An overview of synthesis, properties, and applications. *Materials Research Express*, *10*(6), 062001.
2. Bera, D., Qian, L., Tseng, T. K., & Holloway, P. H. (2010). Quantum dots and their multimodal applications: A review. *Materials*, *3*(4), 2260–2345.
3. Chen, T., Hu, X., Xu, Y., Wang, L., Jiang, W., Jiang, W., & Xie, Z. (2019). Hydrothermal synthesis of highly fluorescent Ag–In–S/ZnS core/shell quantum dots for white light-emitting diodes. *Journal of Alloys and Compounds*, *804*, 119–127.
4. Noro, K., Kozuka, Y., Matsumura, K., Kumasaka, T., Fujiwara, Y., Tsukazaki, A., ... & Otsuka, T. (2024). Parity-independent Kondo effect of correlated electrons in electrostatically defined ZnO quantum dots. *Nature Communications*, *15*(1), 9556.
5. Efros, A., & Efros, AI. L. (1982). Interband absorption of light in a semiconductor sphere. *Soviet Physics: Semiconductors*, *16*(7), 772–775.
6. Ekimov, A. I. (1981). Quantum size effect in three-dimensional microscopic semiconductor crystals. *JETP Letters*, *34*, 345.
7. Brus, L. E. (1984). Electron–electron and electron–hole interactions in small semiconductor crystallites: The size dependence of the lowest excited electronic state. *Journal of Chemical Physics*, *80*(9), 4403–4409.
8. Murray, C. B., Norris, D. J., & Bawendi, M. G. (2005). Synthesis and characterization of nearly monodisperse CdE (E = S, Se, Te) semiconductor nanocrystallites. *SPIE Milestone Series MS*, *180*, 43.
9. Alivisatos, A. P. (1996). Semiconductor clusters, nanocrystals, and quantum dots. *Science*, *271*(5251), 933–937.
10. Klimov, V. I. (2003). Charge carrier dynamics and optical gain in nanocrystal quantum dots: From fundamental photophysics to quantum-dot lasing. In *Semiconductor and Metal Nanocrystals* (pp. 168–221). CRC Press.
11. Singh, P., Kumar, R., & Singh, R. K. (2019). Progress on transition metal-doped ZnO nanoparticles and its application. *Industrial & Engineering Chemistry Research*, *58*(37), 17130–17163.
12. Ozgur, U. (2005). A comprehensive review of ZnO materials and devices. *Journal of Applied Physics*, *98*(41301), 1–103.
13. Baruah, S., & Dutta, J. (2009). Zinc oxide nanoparticles in wastewater treatment. *Water Science and Technology*, *60*(3), 605–618.
14. Fonoberov, V. A., & Balandin, A. A. (2006). ZnO quantum dots: physical properties and optoelectronic applications. *Journal of Nanoelectronics and Optoelectronics*, *1*(1), 19–38.

15. Ca, N. X., Hien, N. T., Fan, X., Do, P. V., Yen, V. H., Hao, P. V., ... & Quang, V. X. (2023). New insights on the luminescence properties and Judd–Ofelt analysis of Er-doped ZnO semiconductor quantum dots. *RSC Advances*, *13*(39), 27292–27302.
16. Forleo, A., Francioso, L., Capone, S., Siciliano, P., Lommens, P., & Hens, Z. (2010). Synthesis and gas sensing properties of ZnO quantum dots. *Sensors and Actuators B: Chemical*, *146*(1), 111–115.
17. Jaffe, J. E., Snyder, J. A., Lin, Z., and Hess, A. C. (2000). LDA and GGA calculations for high-pressure phase transitions in ZnO and MgO. *Physical Review B*, 62, 1660.
18. OpenAI. (2024). Schematic illustration of the 'particle in a box' model for a quantum dot. https://chatgpt.com/g/g-p-676d45f118408191a1404c0b82007dd3-research-paper/c/6772cf9d-24e4-8013-b828-5a6fef6d6b3b
19. Vogel, D., Krüger, P., and Pollmann, J. (1995). Ab initio electronic-structure calculations for II-VI semiconductors using self-interaction-corrected pseudopotentials. *Physical Reviews B*. 52, R14316.
20. Wang, Z. L. (2004). Zinc oxide nanostructures: growth, properties and applications. *Journal of Physics: Condensed Matter*, *16*(25), R829.
21. Wang, Z. L., Kong, X. Y., & Zuo, J. M. (2003). Induced growth of asymmetric nanocantilever arrays on polar surfaces. *Physical Review Letters*, *91*(18), 185502.
22. OpenAI. (2024). A flowchart detailing the sol–gel synthesis process of ZnO quantum dots. https://chatgpt.com/g/g-p-676d45f118408191a1404c0b82007dd3-research-paper/c/6772cf9d-24e4-8013-b828-5a6fef6d6b3b
23. Janotti, A., & Van de Walle, C. G. (2009). Fundamentals of zinc oxide as a semiconductor. *Reports on Progress in Physics*, *72*(12), 126501.
24. Jędrzejewska, M., Wolska-Pietkiewicz, M., Drużyński, Z., & Lewiński, J. (2023). Organometallic one-pot synthesis of ZnO quantum dots coated by sulfoxides as L-type ligands. *Journal of Materials Chemistry C*, *11*(43), 15016–15029.
25. Hohenberg, P., & Kohn, W. (1964). Inhomogeneous electron gas. *Physical Review, 136*(3B), B864.
26. OpenAI. (2024). Flowchart showing the DFT workflow, from input structure to bandgap calculation. https://chatgpt.com/s/t_684ef58d9c7481918ae296f4934b806c
27. Runge, E., & Gross, E. K. U. (1984). Time-dependent density-functional theory. *Physical Review Letters, 52*(12), 997–1000.
28. Brus, L. E. (1984). Electron–electron and electron-hole interactions in small semiconductor crystallites: The size dependence of the lowest excited electronic state. *Journal of Chemical Physics*, *80*(9), 4403–4409.
29. Wikipedia contributors. Piezoelectricity. Wikipedia, the free encyclopedia. [Online]. https://en.wikipedia.org/wiki/Piezoelectricity.
30. Zhang, Y., Leng, Y., Willatzen, M., & Huang, B. (2018). Theory of piezotronics and piezo-phototronics. *Mrs Bulletin*, *43*(12), 928–935.
31. Zhou, Z. Q., Taylor, K., Gjonaj, E., Frömling, T., & Xu, B. X. (2019). Finite element simulations on piezoelectric modulation of ZnO grain boundary barrier height. *Journal of Applied Physics*, *126*(20).

32. Das, P. *Phonon Dispersion*. Quantum Espresso Tutorial. [Online]. https://pranabdas.github.io/espresso/hands-on/phonon/
33. Lee, H. K., Park, T., & Yoo, H. (2024). Device applications enabled by bandgap engineering through quantum dot tuning: A review. *Materials*, *17*(21), 5335.
34. Vanderbilt, D. (1990). Soft self-consistent pseudopotentials in a generalized eigenvalue formalism. *Physical Review B*, *41*(11), 7892.
35. Zhou, Z. Q., Taylor, K., Gjonaj, E., Frömling, T., & Xu, B. X. (2019). Finite element simulations on piezoelectric modulation of ZnO grain boundary barrier height. *Journal of Applied Physics*, *126*(20).
36. Muşat, V., Tăbăcaru, A., Vasile, B. Ş., & Surdu, V. A. (2014). Size-dependent photoluminescence of zinc oxide quantum dots through organosilane functionalization. *RSC Advances*, *4*(108), 63128–63136.
37. Lee, H. K., Park, T., & Yoo, H. (2023). Quantum dot applications in bandgap engineering. *Materials*, *17*(21), 5335. www.mdpi.com/1996-1944/17/21/5335
38. Lee, H. K., Park, T., & Yoo, H. (2024). Device applications enabled by bandgap engineering through quantum dot tuning: A review. *Materials*, *17*(21), 5335.
39. Quantum dot tunability in advanced devices. Howard University SharePoint Document. [Online]. https://howardumy.sharepoint.com/:w:/r/personal/wisdom_benson_bison_howard_edu/_layouts/15/Doc.aspx?sourcedoc=%7B962895ce-f97a-4785-8cf927f268c484ee%7D&action=view&wdAccPdf=0&wdparaid=508DC871
40. Wikipedia contributors. Quantum-confined Stark effect. Wikipedia, the free encyclopedia. https://en.wikipedia.org/wiki/Quantum-confined_Stark_effect
41. Wikipedia contributors. Zeeman effect. Wikipedia, the free encyclopedia. https://en.wikipedia.org/wiki/Zeeman_effect
42. Wen, G. W., Lin, J. Y., Jiang, H. X., & Chen, Z. (1995). Quantum-confined Stark effects in semiconductor quantum dots. *Physical Review B*, *52*(8), 5913.
43. Scappucci, G. (2021). Semiconductor materials stacks for quantum dot spin qubits. arXiv preprint arXiv:2102.10897.
44. Singh, P., Singh, R. K., & Kumar, R. (2021). Journey of ZnO quantum dots from undoped to rare-earth and transition metal-doped and their applications. *RSC Advances*, *11*(4), 2512–2545.
45. Ning, J. Q., Zheng, C. C., Zhang, X. H., & Xu, S. J. (2015). Strong quantum confinement effect and reduced Fröhlich exciton–phonon coupling in ZnO quantum dots embedded inside a SiO_2 matrix. *Nanoscale*, 7(41), 17482–17487.
46. Noro, K., Kozuka, Y., Matsumura, K., Kumasaka, T., Fujiwara, Y., Tsukazaki, A., ... & Otsuka, T. (2024). Parity-independent Kondo effect of correlated electrons in electrostatically defined ZnO quantum dots. *Nature Communications*, *15*(1), 9556.
47. Burkard, G., Ladd, T. D., Pan, A., Nichol, J. M., & Petta, J. R. (2023). Semiconductor spin qubits. *Reviews of Modern Physics*, *95*(2), 025003.

48. Xue, X., Russ, M., Samkharadze, N., Undseth, B., Sammak, A., Scappucci, G., & Vandersypen, L. M. (2022). Quantum logic with spin qubits crossing the surface code threshold. *Nature*, *601*(7893), 343–347.
49. Lilieholm, J. F., Niaouris, V., Kato, A., Fu, K. M. C., & Blinov, B. B. (2020). Photon-mediated entanglement scheme between a ZnO semiconductor defect and a trapped Yb ion. *Applied Physics Letters*, *117*(15).
50. Zhong, S. (2023). Quantum entanglement: An analysis via the orthogonality relation. *Foundations of Physics*, *53*(4), 75.
51. Yatsui, T., Ryu, Y., Morishima, T., Nomura, W., Kawazoe, T., Yonezawa, T., … & Ohtsu, M. (2010). Self-assembly method of linearly aligning ZnO quantum dots for a nanophotonic signal transmission device. *Applied Physics Letters*, *96*(13), 133106, 1–3.
52. Chakraborty, T., Manaselyan, A., & Barseghyan, M. (2018). Effective tuning of electron charge and spin distribution in a dot-ring nanostructure at the ZnO interface. *Physica E: Low-Dimensional Systems and Nanostructures*, *99*, 63–66.
53. Mora-Ramos, M. E., Vinasco, J. A., Laroze, D., Radu, A., Restrepo, R. L., Heyn, C., … & Duque, C. A. (2021). Electronic structure of vertically coupled quantum dot-ring heterostructures under applied electromagnetic probes. A finite-element approach. *Scientific Reports*, *11*(1), 4015.
54. Biborski, A., Kądzielawa, A. P., Gorczyca-Goraj, A., Zipper, E., Maśka, M. M., & Spałek, J. (2016). Dot-ring nanostructure: Rigorous analysis of many-electron effects. *Scientific Reports*, *6*(1), 29887.
55. Flores, P. C., Pablico, D. A. L., & Galapon, E. A. (2024). Partial- and full-tunneling processes across potential barriers. *Europhysics Letters*, *145*(6), 65002.
56. Alexandrov, A., Zvaigzne, M., Lypenko, D., Nabiev, I., & Samokhvalov, P. (2020). Al-, Ga-, Mg-, or Li-doped zinc oxide nanoparticles as electron transport layers for quantum dot light-emitting diodes. *Scientific Reports*, *10*(1), 7496.
57. Tang, J. F., Sie, Y. D., Tseng, Z. L., Lin, J. H., Chen, L. C., & Hsu, C. L. (2022). Perovskite quantum dot–ZnO nanowire composites for ultraviolet–visible photodetectors. *ACS Applied Nano Materials*, *5*(5), 7237–7245.
58. Park, D. S., Wang, H., Vasheghani Farahani, S. K., Walker, M., Bhatnagar, A., Seghier, D., … & McConville, C. F. (2016). Surface passivation of semiconducting oxides by self-assembled nanoparticles. *Scientific Reports*, *6*(1), 18449.

Chapter 9

Advanced device applications of nanomaterials in diagnostic and therapeutic tools

Francis Mensah and Briana Velinor

9.1 INTRODUCTION AND APPLICATIONS

9.1.1 Computed tomography

Computed tomography (CT) (Figure 9.1) is a computerized imaging procedure in which x-rays are projected through a patient at different angles. Narrow beams of x-rays are used by aiming at a specific part of the body and quickly rotating to create detailed images such as cross sections or slices. These images can then be used to help doctors diagnose injuries and diseases. The quality of the CT image can be improved by injecting radiopaque compounds into the body to create definition in vascular structures making it easier to differentiate. There are many advantages to CT because they are fast, accurate, painless, affordable, non-invasive, and widely available. The only disadvantage is that it involves exposure to ionizing radiation. Although it can be minimized by carefully selecting the image parameter for the regions of interest, the risk is still evident. Iodine is one of the primary elements used in CT, but researchers are working to find other alternative radio dense elements because of the side effects caused by the iodinated contrast agent.

9.1.2 Computed tomography applications

CT can be used for observation and analyses of various organs. Thus, we have abdominal and pelvic CT, dental cone beam CT, body CT, discography (discogram), cardiac CT for calcium scoring, facet joint block, chest CT, head CT, coronary CT angiography (CTA), kidney biopsy, CTA, parathyroid 4D CT scan, CT colonography, pediatric CT, CT enterography, positron emission tomography (PET)/CT, CT perfusion imaging of the head, radiation dose, CT safety during pregnancy, spine CT, CT sinuses, urography.

 DOI: 10.1201/9781003512899-9

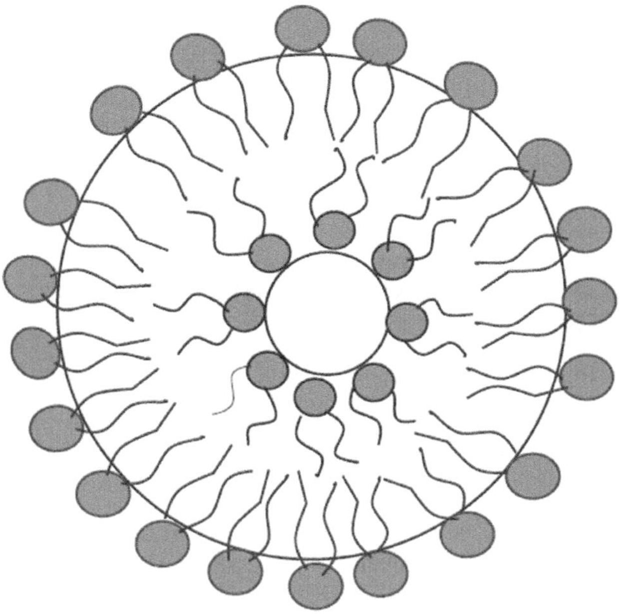

Figure 9.1 Picture of a computed tomography.

9.1.3 Ultrasound

An ultrasound is an imaging test that uses reflected sound waves with frequencies above 20 MHz to build images. This technique has been used for several years to detect anatomic features in the human body. Ultrasound is used for purposes such as cardiac abnormalities and blood vessel wall inflammation which can lead to heart disease and blood clots. Ultrasounds are used for not only detection but also therapeutic purposes. Compared to other imaging techniques, ultrasound has a greater advantage. It is easily portable, inexpensive, and has no risk of exposure to ionizing radiation. It also does not require the use of agents such as iodine or gadolinium. This causes it to have a high resolution over magnetic resonance imaging (MRI).

The ultrasound resolution of the order of tens of microns allows the imaging of small blood vessels. One of the only disadvantages of ultrasound is that the imaging may be confined to a small specific area. In recent years, there has been some improvement in ultrasound. The introduction of ultrasound agents has created differences in the sound reflection or echogenicity between different anatomic compartments. Similarly, with CT, ultrasound contrast agents (USCs) possess the characteristics of having low toxicity and ease of formulation and administration. USCs can achieve a good signal-to-noise ratio. However, the nature of imaging modality brings additional constraints to CT contrast agents. An example is large bubbles: even though they are required for good echogenicity if they are too large that they can be easily trapped in the microvasculature, especially in the lung. The current microbubbles that are used are less than 5 μm in diameter. They must be flexible to produce signals. They must also be strong enough to experience the pressures on the left side of the heart without being ruptured even after long periods of time.

9.2 DEVELOPMENT OF ULTRASOUND CONTRAST AGENTS

Three different components are important in the development of UCAs: the shell, the core, and the targeting moiety.

9.2.1 The shell

The physical property of the shell is very important. The elasticity of the shell will affect the production of a signal. If the shell is too elastic, it may not produce a signal. To obtain a UCA with optimal results, a brittle shell may indicate particularly as an agent for drug deliveries that can be ruptured at the target site. The shell is made of protein, lipid, galactose, or other polymers. In general, lipid-based shells are advantageous because the size (<5 μm) is controllable and their composition can be altered to optimize their properties. It is important to recall that liposomes play a key role in the production of the shells and echogenicity. It is proved that a high concentration of phosphatidylethanolamine (PE) [about 4%], low concentration of phosphatidylglycerol (PG) [about 1%], and constant cholesterol (30%) yielded highest echogenicity [1]. Thus, echogenicity is related to the structure of the liposome.

9.2.2 The core

The fluid entrapped with the microbubbles affects its imaging characteristics. Additionally, air can get into the microbubbles, but because air is soluble in blood, it quickly escapes from the bubble and reduces the potential imaging time. However, the rate of diffusion of air from the microbubbles can be reduced by biopolymer shells. Alkan-Onyuksel, H et al. [2] studied the echogenicity of several polyfluorinated carbon gases (PFCs) in the range of temperatures between 25°C and 45°C.

It is known that PFCs are poorly soluble in water and are liquid at room temperature. Their study proves that perfluorohexacene had the lowest acoustic impedance within the range of 25°C–45°C. The study also concluded that perfluorohexacene would serve as an effective enhancement agent as it has a different impedance compared to water, resulting in good contrast. Kleger et al. further their investigation by extending their research over other fields as core of targeted microbubbles in vitro [3]. The goal in this case was to investigate the fact that targeting the microbubbles to a surface improves the impedance. Finally, the authors suggested that the improvement was due to the development of a reflective interface between the targeted surface and its surroundings. In general, the incorporation of fluids into nanoparticles presents a challenge during the time of scanning that can be met in various ways.

9.2.3 The targeting moiety

UCAs can be directed to target specific tissues in two ways: by inclusion of a targeting moiety on the chemical constituent of the UCA or by a chemical attachment on a ligand (a peptide or an antibody) to one of its constituents. The size and the composition of the shell affect the distribution in vivo, providing some good biological information. This is a passive targeting that can inform on blood flow, the lymphatic system, or other systems. One can argue that microbubbles with shell, which consist of denatured albumin, represent both passive and active targeting of inflammation. It has been noted that, in some cases, albumin-based microbubbles are retained in the myocardium, leading to enhanced ultrasound signal. It has been hypothesized that this was due to an interaction between the microbubbles and the inflamed endothelial tissue. Villanueva et al. [4] tested and proved this hypothesis. They found that the microbubbles adhere strongly to the exposed inflamed extracellular matrix of endothelial cells.

9.2.4 Obstetrical ultrasound

Obstetrical ultrasound is a useful clinical test to explore the presence of a living embryo/fetus and to determine the age of a pregnancy. It also helps to diagnose congenital abnormalities of the fetus and to evaluate the positions of the fetus and the placenta. It helps in many other aspects such as to evaluate the amount of amniotic liquid around the baby and to assess fetal growth and well-being.

9.3 LIPOSOMES AS DIAGNOSTIC AND THERAPEUTIC TOOLS

Liposomes (Figure 9.2) are tiny spherules encompassing concentric aqueous compartments separated from each other and separated from the aqueous surrounding environment by closed layers of amphipathic lipids. In most cases, the lipids are phospholipids. They are oriented like the lipids in bio membranes with hydrophilic head polar groups facing outward compartments and hydrophobic segments that are not polar facing inwards out of contact with water. The ability of liposomes to be used as diagnostic

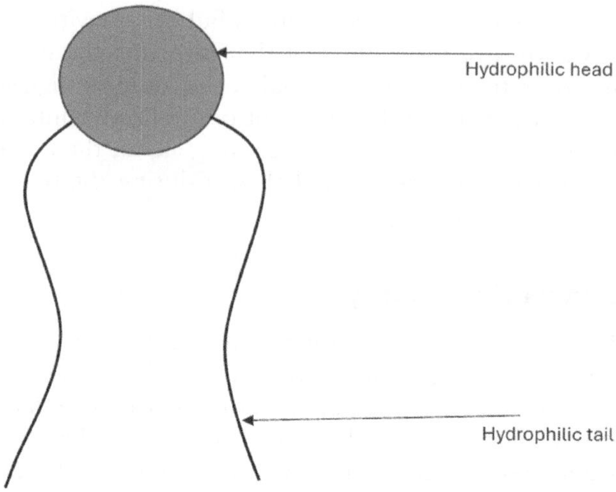

Figure 9.2 Schematic of a liposome with the hydrophilic head and the hydrophobic tail.

tools depends on their ability to entrap a detectable marker molecule within their cavities. For in vivo detection, the marker can be a gamma (γ)-emitting radionucleotide, which can be detected by appropriate devices such as gamma camera or a rectilinear scanner [5]. Liposomes, also known as vesicles, have been used in many cases in vitro for detection. Liposomes can entrap various water-soluble compounds within the inner aqueous solution. They can also entrap lipophilic agents between liposomal bilayers. All these have made them a unique tool for drug delivery and for carrying diagnostic agents for all imaging modalities, such as scintigraphy, MRI, sonography, and CT. Levchenko et al. studied liposomes in diagnostics and treatment of cardiovascular disorders [6,7]. They found that myocardial infarcts can be visualized by sonography when gas-filled liposomes are used for infant imaging in rabbits. Liposomes can be used not only for drug delivery, but also for gene delivery. It is also used for diagnosis of thrombosis and monitoring thrombolytic therapy which can be done and immune-liposomes. It has been proved that nanocarriers for thrombolytic drug deliveries have been efficient to decrease ischemic damage after thrombus formation. One of the very first thrombolytic agents used was heparin which is taken orally and did not prove to be efficient. However, liposomal heparin administered to rats intravenously proved to stay longer in plasma than when taken orally. Inhalable liposomal heparin was later introduced in the rat model of pulmonary embolism and deep vein thrombosis. Inhaled doses of PEGylated liposomes loaded with low-molecular-weight heparin proved to be efficient in reducing thrombus weight. Similarly, efficient results were obtained with a once-a-day dose of subcutaneously administered drug. Furthermore, liposomal prostaglandin E, administered intravenously before thrombolytic therapy, showed a significant shortening of thrombolysis time, a reduction of size, and an improvement in coronary patency and blood flow during reperfusion.

9.4 CONCLUSION

The development of 1D or 2D nanoscale devices has improved observation, detection, and therapeutic actions in biology and nanomedicine. Some of these devices can be used as carriers for drug transport to target cancer cells. Several types of carriers such as macromolecules, cells, viruses, and synthetic particles have proven to be useful. In years to come, drug carriers such as liposomes should continue to play an important role in optimizing drug actions.

ACKNOWLEDGMENT

I would like to thank Ms. Briana Velinor for her assistance in typing part of the draft of this paper.

REFERENCES

[1] Alkan-Onyuksel, H., Demos, S. M., Lanza, G. M., Vonesh, M. J., Klegerman, M. J. Kane, B J. Kuszak, J. McPherson, D. D., Development of inherently echogenic liposomes as an ultrasonic contrast agent. *J Pharm Sci*, 1996, 85(5):486–490.

[2] Keller, G. In vitro differentiation of embryonic stem cells. *Curr Opin Cell Biol*, 1995, 7(6):862–869.

[3] Kleger, A., Sufferlein, T, Malan, D., Tischendorf, M, Storch, A, Wolheim, A., et al. Modulation of calcium-activated potassium channels induces cardio genesis of pluripotent stem cells and enrichment of pacemaker-like cells. *Circulation*, 2010, 122(18):1823–1836.

[4] Villanueva, F.S., et al., Albumin microbubbles adherence to human coronary endothelium: Implications for assessment of endothelial function using myocardial contrast echocardiography. *J Am Coll Cardiol*, 1997, 30(3):689–693.

[5] Gregoriadis, G., Allison, A.C., (eds.) Liposomes as Diagnostic Tools, I. Ross Mc Dougall, Liposomes in Biological Systems 1980, John Willey & Sons, Ltd.

[6] Levchenko, T., Hartner, W., Torchilin, V., Liposomes in diagnostic and treatment of cardiovascular disorders. *Methodist Debakey Cardiovasc J*, 2012, 8(1):36–41.

[7] Woodberry, A., Mensah, F., Applications of nanoparticles in nanomedicine. *Journal Spectroscopy and Characterization of Nanomaterials and Novel Materials: Experiments, Modeling, Simulations, and Applications*. 2022:345–360. Wiley-VCH GmbH.

Chapter 10

Field-effect transistors

Advances, challenges, and future prospects

Priyanka U. Londhe and Nandu B. Chaure

10.1 INTRODUCTION TO NANOELECTRONICS AND FIELD-EFFECT TRANSISTORS (FETS)

10.1.1 Overview and basics

Nanoelectronics is a revolutionary field within electronics that focuses on devices and systems operating at the nanometer scale, typically less than 100 nm. By utilizing principles of quantum mechanics and nanotechnology, nanoelectronics enables unprecedented advancements in size, speed, efficiency, and functionality. This domain has emerged as a crucial enabler of modern technologies, driving innovations in computing, communication, healthcare, and energy systems.

At its core, nanoelectronics involves designing and fabricating electronic components using nanoscale materials and structures. Unlike conventional microelectronics, where devices operate based on bulk material properties, nanoelectronics exploits quantum effects, such as electron tunneling and energy quantization, to achieve enhanced performance. These effects become prominent at the nanoscale, making nanoelectronics distinct and revolutionary. Components such as transistors, memory devices, sensors, and light-emitting diodes (LEDs) have undergone significant transformations through the application of nanotechnology principles.

One of the foundational aspects of nanoelectronics is miniaturization. As transistor dimensions shrink, they offer increased performance and energy efficiency while enabling higher device densities on a single chip. This has led to the evolution of Moore's Law, which predicts the doubling of transistors on an integrated circuit (IC) every 2 years. Although traditional silicon-based electronics face physical and material limitations at nanoscale dimensions, nanoelectronics introduces alternative materials and architectures to sustain this trend. For instance, nanowires, carbon nanotubes (CNTs), and two-dimensional (2D) materials like graphene and molybdenum disulfide (MoS_2) are being explored to replace or augment silicon in future devices.

 DOI: 10.1201/9781003512899-10

Quantum mechanics plays a vital role in nanoelectronics. At nanoscale dimensions, electrons behave as waves rather than particles, leading to phenomena like quantum tunneling, quantized conductance, and electron confinement. These effects are exploited in devices such as quantum dots, single-electron transistors (SETs), and spintronic components. For example, quantum dots, which are nanoscale semiconductor particles, are used in displays, solar cells, and medical imaging due to their tunable optical and electronic properties.

The development of nanoelectronics also hinges on the discovery and synthesis of advanced materials. Carbon-based materials like graphene and CNTs have garnered significant attention due to their exceptional electrical conductivity, mechanical strength, and flexibility. In addition, transition metal dichalcogenides (TMDs) such as MoS_2 exhibit semiconducting properties, making them ideal candidates for next-generation transistors and optoelectronic devices. These materials offer the potential to overcome the limitations of silicon, enabling the creation of smaller, faster, and more energy-efficient components.

Fabrication techniques play a crucial role in the advancement of nanoelectronics. Methods such as atomic layer deposition (ALD), electron-beam lithography, and molecular self-assembly allow for precise control at the atomic level, enabling the construction of nanoscale devices. Though complex and resource-intensive, these techniques are essential for achieving the desired performance and reliability in nanoelectronic systems. Furthermore, emerging approaches like bottom-up nanofabrication, which relies on the self-assembly of nanostructures, hold promise for scalable and cost-effective manufacturing.

Nanoelectronics has found applications across diverse fields. In computing, it enables the development of ultra-fast processors, high-capacity memory devices, and neuromorphic systems that mimic the human brain's functionality. In healthcare, nanoelectronic sensors are used for early disease detection, personalized medicine, and implantable devices. The field also contributes to renewable energy solutions, such as high-efficiency solar cells and energy storage devices, and supports advancements in communication technologies, including 5G and beyond.

Despite its transformative potential, nanoelectronics faces several challenges. Fabricating devices at the nanoscale demands precise and expensive techniques, while maintaining reliability and performance over time is complex. Managing heat dissipation, avoiding quantum noise, and addressing material stability are additional hurdles. Nonetheless, ongoing research continues to address these issues, paving the way for further advancements.

Nanoelectronics represents the next frontier in electronics, bridging the gap between classical and quantum systems. It offers the tools to redefine modern technology, enabling innovations that are smaller, faster, and more efficient. As research in materials, fabrication, and device design progresses, nanoelectronics will continue to shape the future of electronics, ensuring its impact across industries and society.

10.1.2 Historical development of field-effect transistors

The FET is one of the most transformative innovations in electronics, forming the backbone of modern devices and ICs. Its conceptual roots date back to the early 20th century, when researchers explored the principles of semiconductor behavior and electronic amplification. Austrian physicist *Julius Edgar Lilienfeld* first proposed the idea of a transistor-like device in 1926, describing how an electric field could control current flow through a semiconductor. Despite this visionary concept, practical implementation was hindered by the lack of suitable materials and fabrication technologies. In 1934, *Oskar Heil* refined these ideas by proposing another device based on field effects, but similar obstacles prevented its realization (Fei, 2024).

Progress in FET development stalled for several decades, as vacuum tubes dominated electronics. The turning point came in 1947 when *John Bardeen*, *Walter Brattain*, and *William Shockley* at Bell Laboratories invented the bipolar junction transistor (BJT). Although their original goal was to develop a working FET, challenges with surface states in semiconductors led them to discover the point-contact transistor instead. This invention revolutionized electronics, laying the groundwork for further semiconductor research and ultimately reigniting interest in FETs (Dash & Jena, 2024).

The first practical FET, the junction field-effect transistor (JFET), was patented by *Heinrich Welker* in 1945. This innovation marked a critical milestone, enabling more controlled current flow using a reverse-biased junction. By 1950, Japanese engineers *Jun-ichi Nishizawa* and *Y. Watanabe* introduced the static induction transistor (SIT), a high-frequency variant of the JFET, which expanded the potential applications of FET technology (Xu et al., 2024).

The most significant breakthrough came in 1960 when *Mohamed Atalla* and *Dawon Kahng* at *Bell Labs* developed the first metal-oxide-semiconductor field-effect transistor (MOSFET). This device utilized a silicon dioxide layer to insulate the gate terminal from the channel, enabling precise control of current with minimal power consumption. The MOSFET's simple design and scalability addressed many of the challenges faced by earlier FETs, marking the beginning of the modern era of transistors. Its ability to integrate into complex circuits quickly made it the dominant transistor type in digital electronics (Kalra, Kannan, & Kumar, 2024).

Throughout the 1970s and 1980s, MOSFETs became central to the rapid development of microprocessors, memory chips, and digital systems. Innovations in lithography and fabrication processes allowed for exponential scaling, aligning with Moore's Law, which predicted the doubling of transistors on a chip every 2 years. This miniaturization enabled faster, more efficient, and cost-effective devices. The introduction of complementary MOS (CMOS) technology further enhanced the power efficiency and performance of ICs, solidifying the MOSFET's role in computing and telecommunications (Saini & Tyagi, 2024).

As the 20th century drew to a close, the limitations of traditional planar MOSFETs became apparent, particularly as device dimensions approached the nanoscale. To address these challenges, Chenming Hu and his team developed the FinFET in the 1990s. This three-dimensional (3D) transistor structure improved gate control, reduced leakage currents, and enhanced performance, paving the way for continued scaling in nanoelectronics. The FinFET's success has made it a cornerstone of modern semiconductor technology, especially in advanced processors (Mohanbabu & Maheswari, 2024).

The evolution of FETs continues today with the exploration of new materials and architectures. Nanowire FETs, carbon nanotube FETs (CNTFETs), and 2D material-based FETs offer promising solutions to the challenges of miniaturization. CNTs, with their high carrier mobility and ballistic transport properties, enable ultra-fast switching and low-power operation. Similarly, 2D materials like graphene and TMDs allow for the creation of ultra-thin, flexible FETs with exceptional electronic properties (Shah & Kashyap, 2024).

FETs have also found applications beyond traditional computing, including in memory devices, sensors, and emerging technologies like quantum and neuromorphic computing. Floating-gate MOSFETs have enabled non-volatile memory technologies, such as flash memory, while SETs and neuromorphic FETs are paving the way for quantum computing and brain-inspired architectures (Yang et al., 2024).

The historical development of FETs reflects a journey from theoretical concepts to practical applications, driven by continuous advancements in materials and fabrication techniques. From Lilienfeld's early patents to the billions of MOSFETs manufactured each year, FETs have become indispensable in shaping the digital world. Their evolution underscores the importance of innovation and collaboration, ensuring that FET technology remains at the forefront of electronics, enabling the next generation of devices and systems (Mo et al., 2025).

10.1.3 Importance of FET in nanoelectronics

FETs are fundamental to the field of nanoelectronics, serving as the cornerstone of modern electronic devices and ICs. Their ability to control electrical current flow through a channel using an electric field makes them versatile and efficient. As the electronics industry shifts toward the nanoscale, FETs continue to evolve, playing a critical role in enabling faster, smaller, and more energy-efficient devices.

The significance of FETs in nanoelectronics lies in their adaptability to scaling. Traditional silicon-based MOSFETs have driven decades of technological advancement, but their physical limitations become evident as dimensions shrink to the nanoscale. Nanoelectronic FETs address these challenges by incorporating advanced materials, architectures, and quantum principles. Innovations such as FinFETs, nanowire FETs, and CNTFETs offer improved performance and scalability, ensuring the continued progress of semiconductor technology.

One of the most critical roles of FETs in nanoelectronics is their impact on power efficiency. As devices become smaller and more densely packed, reducing power consumption becomes essential to mitigate heat generation and enhance battery life in portable applications. FETs designed with high-k dielectrics, low-dimensional materials like graphene, or TMDs significantly lower leakage currents and operating voltages, addressing these issues. For instance, 2D materials enable the creation of ultra-thin FETs with enhanced electron mobility, reducing both power loss and heat dissipation.

FETs are also pivotal in maintaining the speed and performance of nanoelectronic devices. With shrinking dimensions, short-channel effects such as threshold voltage roll-off and drain-induced barrier lowering (DIBL) pose challenges. Advanced FET architectures like FinFETs and gate-all-around (GAA) FETs overcome these issues by enhancing gate control over the channel. These designs minimize performance degradation, enabling transistors to operate at higher speeds with greater efficiency, a necessity for high-performance computing and data processing.

The versatility of FETs extends to their material compatibility, which is essential in nanoelectronics. Beyond traditional silicon, materials like CNTs, graphene, and MoS_2 have been integrated into FET designs, leading to improved switching speeds, higher current densities, and tunable electronic properties. CNTFETs, for example, exhibit ballistic transport properties that allow electrons to move with minimal scattering, significantly boosting transistor performance. Similarly, graphene's high carrier mobility and flexibility have opened pathways for developing transparent, flexible, and wearable electronics.

FETs are indispensable in emerging nanoelectronic applications. In the realm of sensing, FETs are used in nanoscale biosensors and gas detectors due to their high sensitivity to changes in the local environment. These sensors can detect biomolecules, chemical gases, and even environmental pollutants with exceptional precision, making them invaluable in healthcare, environmental monitoring, and industrial processes. In addition, FETs play a critical role in memory devices, enabling advanced non-volatile memory technologies such as resistive RAM (ReRAM) (Chaure et al., 2009) and ferroelectric FETs (FeFETs) (Xu et al., 2024).

In the context of quantum and neuromorphic computing, FETs demonstrate their adaptability to future technologies. SETs, a variant of FETs, exploit quantum tunneling effects to achieve precise control of electron flow at the atomic level, making them suitable for quantum computing. Similarly, FET-based neuromorphic systems mimic the functionality of biological neurons and synapses, paving the way for energy-efficient artificial intelligence and machine learning systems.

The importance of FETs in nanoelectronics is unparalleled. Their ability to adapt to scaling challenges, integrate advanced materials, and support diverse applications ensures their centrality in the future of electronics. As technology continues to evolve, FET innovations will remain at the forefront, enabling the realization of smaller, faster, and more efficient devices that define the next era of nanoelectronics.

10.2 FUNDAMENTALS OF FIELD-EFFECT TRANSISTORS

FETs are unipolar devices, meaning their operation primarily involves one type of charge carrier—either electrons (in n-channel devices) or holes (in p-channel devices). Unlike BJTs, FETs rely on an electric field to control the flow of carriers, offering high input impedance and low power consumption. This makes FETs ideal for use in ICs, where power efficiency is critical. FETs are also preferred in applications requiring minimal input current, such as in sensors and amplifiers, symbols are shown in Figure 10.1.

The core concept of a FET is the control of a conductive channel between two terminals, the source and drain, by a voltage applied to a third terminal called the gate. The gate voltage influences the charge density in the channel, allowing precise control of the current. The lack of direct gate current

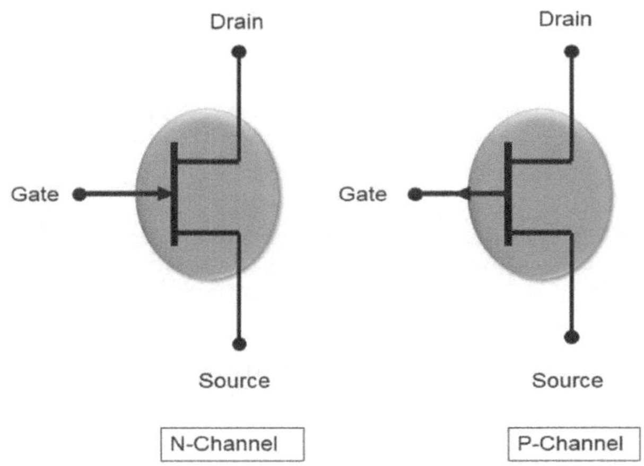

Figure 10.1 FET symbols.

(except leakage) reduces power consumption, which is a significant advantage over BJTs. FETs are classified into various types based on their structure and material composition, each optimized for different applications, ranging from high-speed switching to flexible electronics.

10.2.1 Basic working principle of FETs

The basic operation of a FET can be explained using the concept of an electric field generated by the gate voltage. When a voltage is applied to the gate terminal, it creates an electric field that either attracts or repels charge carriers in the channel (Figure 10.2). The resulting change in carrier density modulates the channel's conductivity, controlling the current flow between the source and drain (Lenka & Nguyen, 2024). In the linear (ohmic) region, the channel behaves like a resistor, and the current increases linearly with the drain–source voltage (VDS). As VDS increases further, the channel reaches a point where it pinches off at the drain end, entering the saturation region.

In this region, the current becomes independent of VDS and depends mainly on the gate–source voltage (VGS). This behavior allows FETs to function as amplifiers and switches (Figure 10.3).

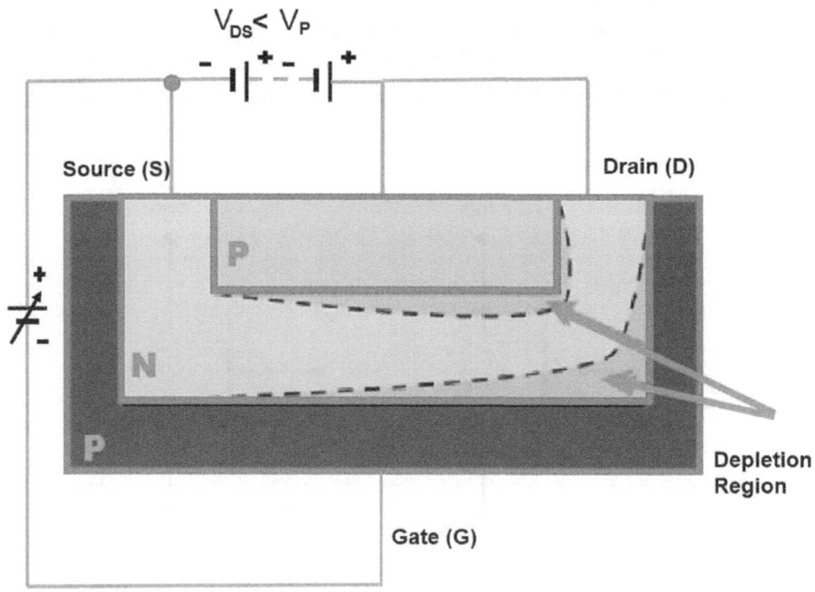

Figure 10.2 Working of FET.

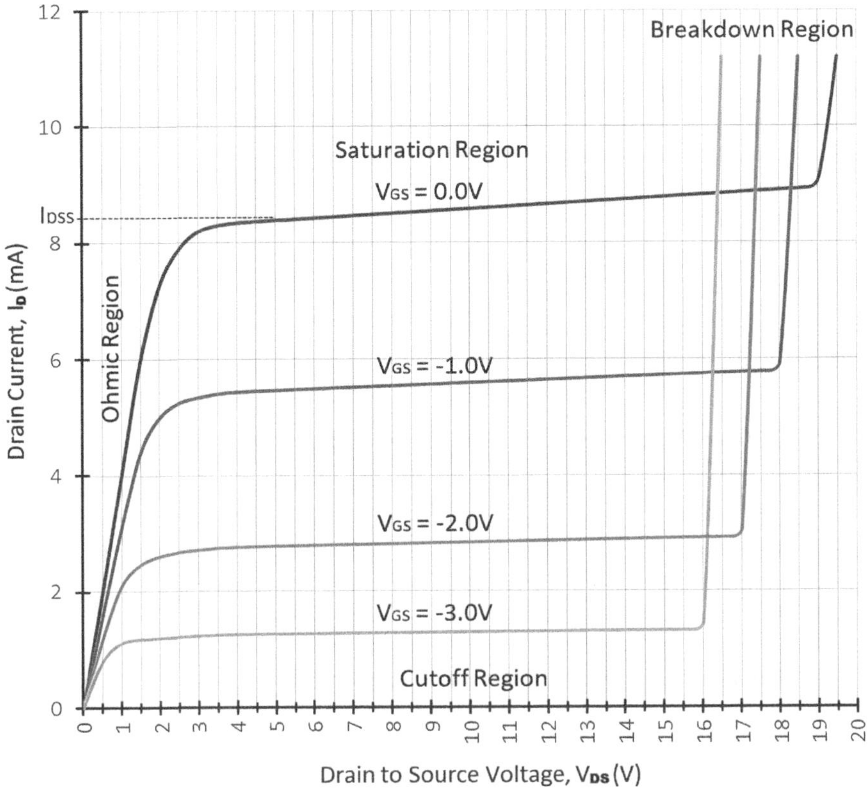

Figure 10.3 I–V characteristics.

10.2.2 Types of FETs

10.2.2.1 Junction field-effect transistor (JFET)

A JFET is the simplest type of FET, where a p–n junction controls the current flow through the channel. The channel can either be n-type or p-type. The gate is reverse-biased relative to the source, creating a depletion region that restricts current flow. The extent of this depletion region depends on the gate voltage (Figure 10.4).

- n-channel JFETs: A negative gate voltage widens the depletion region, reducing current flow.
- p-channel JFETs: A positive gate voltage reduces current flow by expanding the depletion region.

JFETs are valued for their high input impedance and low noise, making them suitable for analog signal amplification. However, their inability to scale well limits their use in modern digital circuits.

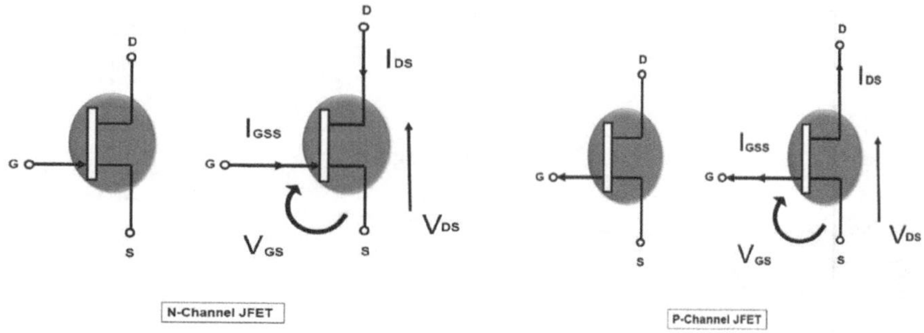

Figure 10.4 JFET symbols.

10.2.2.2 Metal-oxide-semiconductor field-effect transistor (MOSFET)

MOSFETs have a more complex structure, featuring an insulated gate separated from the channel by a thin oxide layer, as shown in Figure 10.5. This insulation allows for extremely high input impedance.

MOSFETs can operate in two modes:

- *Enhancement mode:* The device is normally off, and a gate voltage is needed to create a conductive channel.
- *Depletion mode:* The device is normally on, and a gate voltage depletes the channel of carriers to turn it off.

MOSFETs dominate the semiconductor industry due to their scalability and versatility. They are used in microprocessors, memory chips, and power electronics. The thin oxide layer, however, makes them sensitive to static discharge, requiring careful handling.

10.2.2.3 High electron mobility transistor (HEMT)

HEMTs utilize a heterostructure, typically combining materials like gallium arsenide (GaAs) or gallium nitride (GaN) with aluminum gallium arsenide (AlGaAs), as shown in Figure 10.6. The heterojunction creates a high-mobility channel known as a two-dimensional electron gas (2DEG), where electrons can move with minimal scattering. This high mobility results in faster switching speeds and better performance at high frequencies. HEMTs are ideal for high-frequency applications, including microwave communication, radar systems, and satellite technology. Their low noise figure also makes them suitable for sensitive signal amplification in wireless receivers.

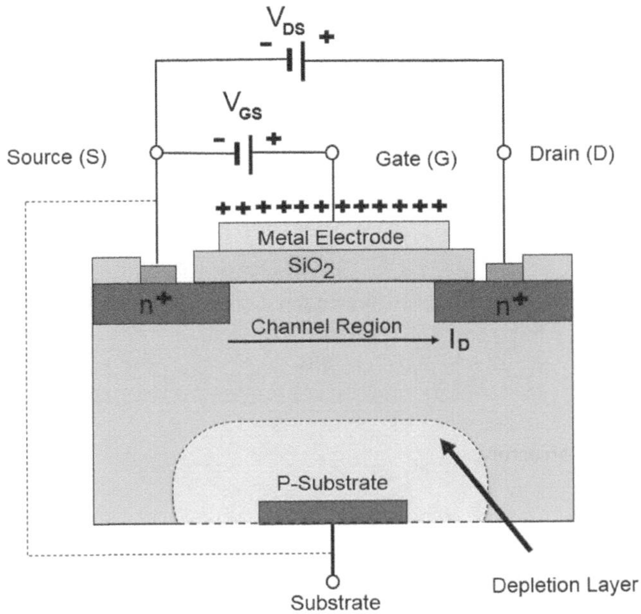

Figure 10.5 Structure of MOSFET.

10.2.2.4 Organic field-effect transistor (OFET)

OFETs are transistors that use organic semiconducting materials, such as conjugated polymers or small molecules; such as entacene, polythiophene, rubrene and phathalocynines (Chaure et al., 2010, 2011) and metal-doped phthalocyanines (Chaure et al., 2012, 2015 and Faris et. al. 2014), as the active channel layer. They consist of a source, drain, and gate electrode, similar to conventional FETs, but are built on flexible or rigid substrates like plastic or glass. The dielectric layer insulates the gate from the organic semiconductor, controlling charge carrier flow. OFETs operate based on the modulation of charge carriers (holes or electrons) in the organic layer when a voltage is applied to the gate. Generally, four common configurations used to fabricate the OFET are schematically represented in Figure 10.7a–d. These transistors offer advantages such as mechanical flexibility, low-cost fabrication, and solution-processability, making them suitable for applications in flexible displays, wearable electronics, and printed circuits. However, they face challenges such as lower charge carrier mobility, environmental stability, and interface defects compared to silicon-based transistors. Advances in material science and device engineering are improving OFET performance, enhancing their commercial viability.

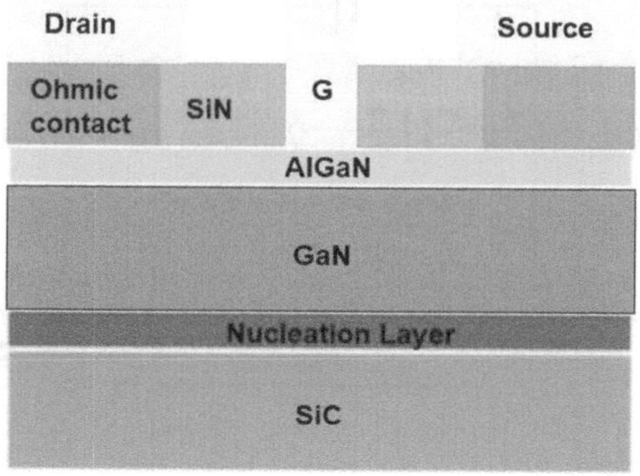

Figure 10.6 HEMT structure.

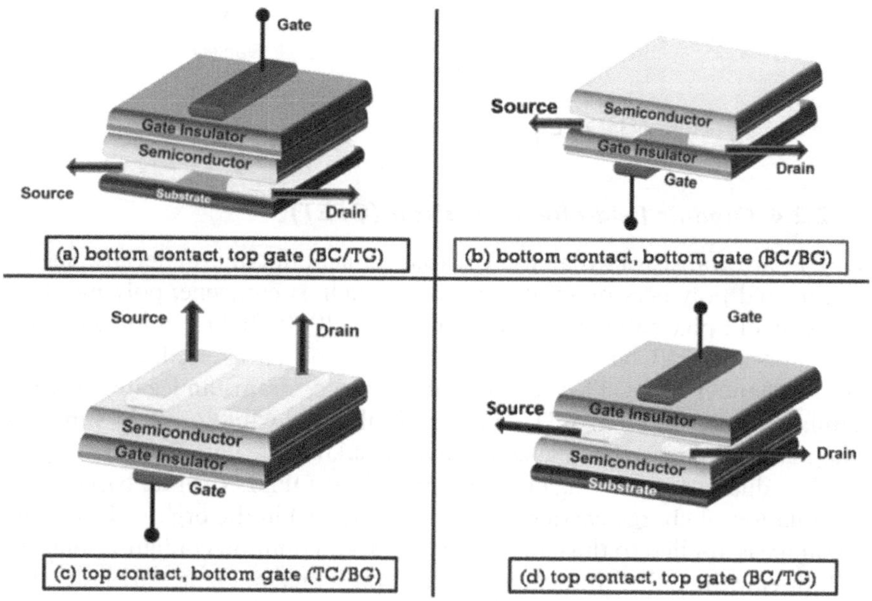

Figure 10.7 Schematic representation of common configurations used for the fabrication of the OFETs.

The use of Self assembled monolayers (SAMs) on the gate oxide layer helps to enhance the field-effect mobility in an afterward grown organic active layer in OFETs (Londhe et al., 2021). Although OFETs offer mechanical flexibility, their performance in terms of mobility and stability is lower than silicon-based FETs. However, ongoing research is improving their characteristics, making them more competitive in niche markets.

10.2.3 Key parameters of FETs

10.2.3.1 Threshold voltage (VTH)

The threshold voltage is a critical parameter that defines the minimum gate-to-source voltage required to turn the FET on. For an enhancement-mode MOSFET, V_{th} is the point where a sufficient number of carriers accumulate in the channel to allow significant current flow. This parameter is influenced by factors such as oxide thickness, channel doping, and temperature. Devices with a low threshold voltage are suited for low-power applications but may suffer from leakage currents.

10.2.3.2 Subthreshold slope

The subthreshold slope measures how effectively a FET can transition from the off state to the on state. It is defined as the change in gate voltage required to change the drain current by one order of magnitude in the subthreshold region (where the device operates below V_{th}). A steep subthreshold slope (low value) is desirable for fast switching and efficient operation. It is typically measured in mV/decade (Lenka & Nguyen, 2024).

10.2.3.3 On-off ratio

The on–off ratio is the ratio of the drain current in the on state (saturation region) to the off state (subthreshold region). A high on–off ratio is essential for digital applications, ensuring a clear distinction between logic states and minimizing power dissipation. For modern MOSFETs, on–off ratios can exceed 10^6, providing excellent switching performance.

10.3 SCALING LAWS IN NANOELECTRONICS

Scaling laws in nanoelectronics govern the miniaturization of transistors and other electronic devices, enabling the exponential growth of computing power over the past few decades. These laws describe how device dimensions, such as channel length, gate oxide thickness, and interconnect widths, can be reduced while maintaining or enhancing performance. The scaling process follows both empirical observations, like Moore's Law, and theoretical guidelines, such as Dennard scaling, which predicts that power density remains constant as transistors shrink (Fei, 2024). However, as devices reach nanometer scales, physical limitations and quantum mechanical effects introduce challenges that require innovative solutions, redefining how scaling continues in modern electronics.

10.3.1 Moore's Law and its impact on FET development

Moore's Law, coined by Gordon Moore in 1965, predicts that the number of transistors on an IC doubles approximately every 2 years, leading to exponential growth in computational power. This principle has driven the continuous scaling of FETs over the decades, pushing manufacturers to innovate smaller, faster, and more energy-efficient devices. As transistors shrink, the cost per function decreases, enhancing performance in fields such as computing, telecommunications, and consumer electronics. Advances in FET technology, particularly in MOSFETs, have been pivotal in meeting Moore's predictions by reducing feature sizes from micrometers to nanometers.

The impact of Moore's Law extends beyond transistor density, influencing the development of new materials, fabrication techniques, and device architectures such as FinFETs and GAA FETs. It has also led to the exploration of alternative devices, including CNTFETs and tunnel FETs (TFETs), to sustain performance gains as silicon scaling approaches its physical limits.

10.3.2 Challenges of scaling down FETs

Scaling FETs to nanometer dimensions introduces several challenges that impact performance, power consumption, and reliability. One major issue is short-channel effects, where the channel length becomes comparable to the depletion region width, reducing the gate's control over the channel. This leads to increased leakage currents and threshold voltage roll-off. In addition, power density rises as devices shrink, leading to significant thermal management challenges.

Gate oxide thinning improves gate control but raises concerns about tunneling currents, which increase power dissipation. The variability in dopant distribution and process imperfections also affects device uniformity and yield. Furthermore, parasitic resistances and capacitances become significant, limiting switching speeds and degrading signal integrity. To address these challenges, new materials such as high-k dielectrics and techniques like strain engineering have been adopted. 3D transistor architectures like FinFETs help mitigate short-channel effects by enhancing electrostatic control over the channel.

10.3.3 Quantum effects in nano-scale FETs

As FETs scale down to the nanometer regime, quantum mechanical effects become prominent, influencing device behavior. Quantum tunneling through thin gate oxides and narrow depletion regions results in significant leakage currents. This phenomenon limits further reduction of oxide thickness and challenges power efficiency targets.

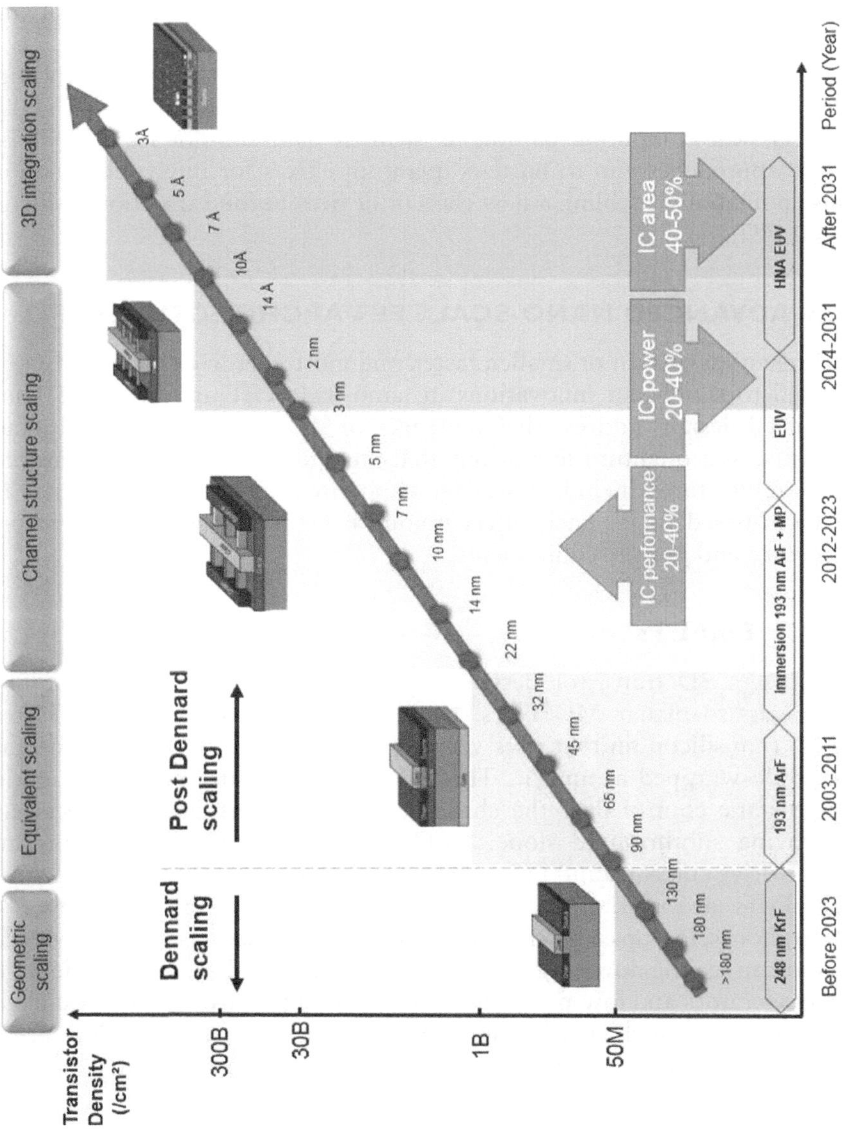

Figure 10.8 Representation of Moore's Law.

Quantum confinement occurs when the channel dimensions approach the electron's de Broglie wavelength, altering the density of states and carrier mobility. This impacts threshold voltage and overall device performance. Additionally, quantum interference and coherence effects may arise in extremely scaled devices, affecting their predictability and performance consistency (Zhang et al., 2024).

To leverage or mitigate these quantum effects, researchers are exploring novel materials like 2D semiconductors (e.g., graphene and TMDs) and non-classical computing paradigms such as quantum-dot FETs (QFETs). These approaches aim to harness quantum effects for improved functionality, potentially enabling a new class of devices beyond traditional silicon-based FETs.

10.4 ADVANCED NANO-SCALE FET ARCHITECTURES

The relentless pursuit of smaller, faster, and more energy-efficient transistors has led to significant innovations in nano-scale FET architectures. These advanced designs address the challenges of short-channel effects, leakage currents, and quantum limitations that emerge at nanometer dimensions. Key architectures include FinFETs, nanowire FETs, CNTFETs, and 2D material-based FETs. Each offers unique benefits tailored to specific performance and power requirements.

10.4.1 FinFETs

FinFETs, a 3D transistor design (Figure 10.9), are the most prominent successors to planar MOSFETs. Unlike traditional designs, FinFETs feature a thin silicon fin that rises vertically from the substrate, with the gate electrode wrapped around it. This multi-gate structure provides superior electrostatic control over the channel, reducing short-channel effects and improving subthreshold slope. FinFETs exhibit lower leakage currents, higher drive currents, and improved scalability down to 7 nm and beyond.

FinFETs are pivotal in modern processors, such as those used in smartphones, laptops, and cloud computing infrastructures. They play a critical role in scaling down logic circuits and enhancing the energy efficiency of AI accelerators and low-power IoT devices. FinFETs are also integral to the development of high-speed memory technologies like Static random access memory (SRAM) and Dynamic random access memory (DRAM).

10.4.2 Nanowire FETs

Nanowire FETs extend the concept of FinFETs by using a GAA configuration. The nanowire channel, which can be cylindrical or rectangular, is completely surrounded by the gate, offering unparalleled electrostatic control,

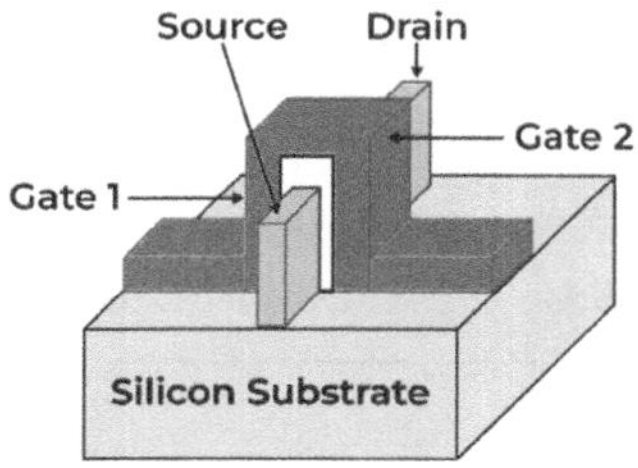

Figure 10.9 Basic structure of FinFET.

as shown in Figure 10.10. This design minimizes leakage and enables even smaller device footprints (Shah & Kashyap, 2024). Nanowire FETs are highly scalable and are considered the primary candidates for technology nodes beyond 5 nm.

Nanowire FETs are being explored for high-density logic circuits and non-volatile memory due to their compact size and enhanced performance. Their GAA structure also makes them ideal for next-generation semiconductor devices requiring ultra-low power and high speed.

10.4.3 Carbon nanotube FETs

CNTFETs utilize CNTs as the channel material instead of silicon. CNTs offer superior electrical properties, including high electron mobility, excellent current-carrying capacity, and low power dissipation. These attributes make CNTFETs a promising technology for ultra-low-power and high-frequency applications (Saini & Tyagi, 2024).

Despite their potential, CNTFETs face challenges such as controlling the chirality of nanotubes (which affects their conductivity) and integrating them with existing silicon processes. However, they are actively researched for use in flexible electronics, next-generation logic devices, and applications requiring extreme energy efficiency (Xu et al., 2024).

10.4.4 2D material-based FETs

2D materials, like graphene and MoS_2, have emerged as attractive options for future FET designs. These materials are just one atom thick, providing an ultra-thin body that minimizes short-channel effects (Mo et al., 2025).

- *Graphene FETs:* Graphene exhibits exceptional carrier mobility and is ideal for high-speed and radio-frequency (RF) applications. However, its lack of a bandgap limits its application in digital electronics.

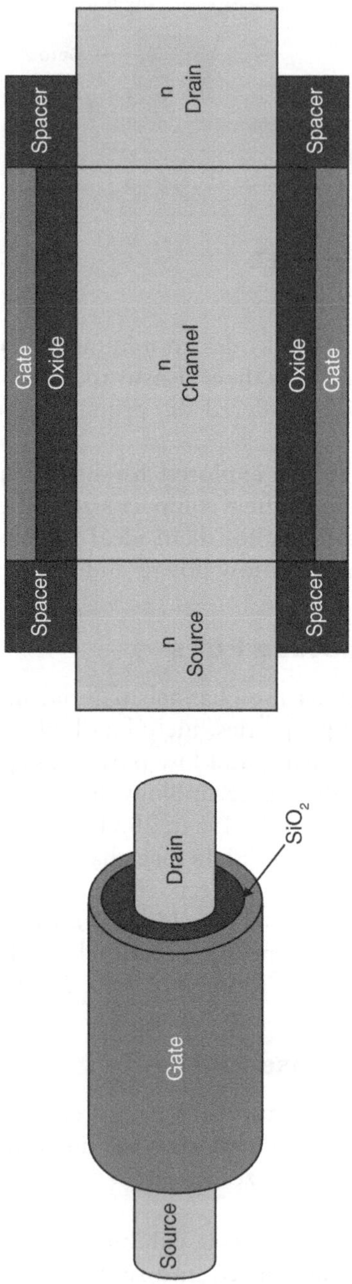

Figure 10.10 A typical circular nanowire FET with its cross-sectional view.

- *MoS_2 FETs:* MoS_2 and other TMDs possess a natural bandgap, making them suitable for low-power applications. MoS_2 FETs are being explored for use in flexible electronics, wearable devices, and sensors (Zhang et al., 2024).

2D material-based FETs are particularly promising in flexible and transparent electronics. They are also critical for integrating sensors and logic circuits into ultra-compact and lightweight form factors, enabling the next generation of wearable and portable technologies (Jiang et al., 2025).

10.5 MATERIALS FOR NANO-FETS

As the dimensions of FETs continue to shrink to the nanoscale, the choice of materials plays a critical role in maintaining device performance, efficiency, and reliability. Traditional silicon-based transistors are reaching fundamental physical and quantum mechanical limits, necessitating the transition to advanced materials that can enable continued progress in nanoelectronics. This chapter explores the evolution from silicon to novel materials, the significance of high-k dielectrics, and the role of emerging semiconducting nanomaterials such as CNTs, graphene, and TMDs in the development of nano-FETs.

10.5.1 Silicon and beyond: Transition to advanced materials

Silicon has been the dominant material in transistor technology for decades due to its excellent electrical properties, mature fabrication techniques, and abundant availability. However, as the semiconductor industry approaches the limits of Moore's Law, new materials are being explored to overcome issues such as short-channel effects, power dissipation, and reduced carrier mobility. Alternative materials, including germanium, compound semiconductors (e.g., GaAs and indium gallium arsenide (InGaAs)), and 2D materials, have demonstrated superior electrical performance in specific applications. These materials offer higher mobility, better electrostatic control, and the potential for low-power operation, making them attractive candidates for next-generation nano-FETs.

10.5.2 High-k dielectrics in nano-FETs

One of the key challenges in nanoscale transistor scaling is maintaining low gate leakage current while ensuring high capacitance for efficient gate control. Traditional silicon dioxide (SiO_2) gate dielectrics suffer from excessive leakage as their thickness is reduced. High-k dielectrics, such as hafnium dioxide (HfO_2), zirconium dioxide (ZrO_2), and aluminum oxide (Al_2O_3), have been introduced to replace SiO_2 in advanced nano-FETs. These materials provide higher capacitance while allowing a physically thicker dielectric layer, thereby reducing tunneling leakage. The integration of high-k dielectrics with novel channel materials also presents challenges in terms of interface defects, reliability, and process compatibility, necessitating further research into interface engineering and material optimization (Al-Turk, 2022).

10.5.3 Role of semiconducting nanomaterials

The shift from bulk silicon to nanoscale materials has led to the exploration of novel semiconducting nanomaterials that exhibit exceptional electronic and structural properties. These materials include carbon-based structures, such as CNTs and graphene, as well as emerging 2D materials like TMDs.

10.5.3.1 Carbon nanotubes

CNTs are cylindrical nanostructures composed of rolled-up graphene sheets. They exhibit remarkable electrical, mechanical, and thermal properties, making them excellent candidates for nano-FETs. Due to their quasi-ballistic transport properties and high carrier mobility, CNT-based transistors can operate at lower power while achieving higher speeds than conventional silicon devices. However, challenges such as chirality control, contact resistance, and large-scale fabrication remain critical hurdles in their commercialization (Tamersit et al., 2022).

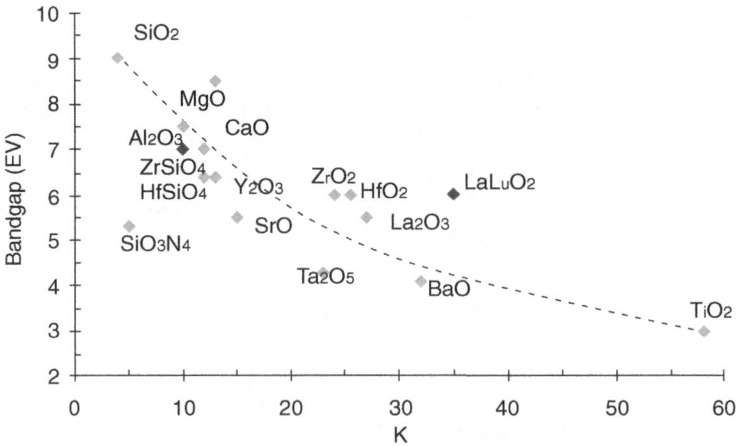

Figure 10.11 High-k dielectric materials in nanoelectronics.

10.5.3.2 Graphene

Graphene, a single layer of carbon atoms arranged in a hexagonal lattice, has attracted significant attention in nanoelectronics due to its ultrahigh carrier mobility and atomic-scale thickness. Unlike silicon, graphene lacks an intrinsic bandgap, which poses a challenge for switching applications in FETs. Various bandgap engineering techniques, such as chemical doping, bilayer graphene manipulation, and patterned nanoribbons, have been investigated to enhance graphene's suitability for transistor applications. Despite these challenges, graphene's superior electrical conductivity and mechanical flexibility make it a promising material for high-speed and flexible electronic devices (Ma, Wang, & Liu, 2024).

10.5.3.3 Transition metal dichalcogenides

TMDs are a class of layered 2D materials with the general formula MX_2, where M represents a transition metal (e.g., Mo, W) and X represents a chalcogen (e.g., S, Se, Te). Unlike graphene, many TMDs, such as MoS_2 and tungsten diselenide (WSe_2), possess a tunable bandgap, making them highly suitable for transistor applications. Their atomically thin nature enables excellent electrostatic control in nano-FETs, reducing short-channel effects. However, challenges such as contact resistance, defect engineering, and large-scale synthesis need to be addressed for practical applications (Radhakrishnan et al., 2023).

In conclusion, the development of advanced nano-FETs relies on the integration of novel materials that offer superior performance over traditional silicon-based transistors. High-k dielectrics and emerging semiconducting nanomaterials, including CNTs, graphene, and TMDs, hold immense potential in enabling future generations of high-performance, low-power electronic devices.

10.6 FABRICATION TECHNIQUES FOR NANO-FETS

Fabricating nano-FETs requires precision and innovation to achieve the miniaturization and performance demanded by modern electronics. Techniques are broadly classified into top-down and bottom-up approaches, each with its own methodologies, applications, and challenges. Advances in lithography and process optimization have played pivotal roles in achieving scalability and efficiency.

10.6.1 Top-down approaches

Top-down approaches are fabrication techniques that start with a bulk material (usually a wafer) and involve the removal or modification of material to create nano-scale features. This method is foundational in nanoelectronics, offering high precision and scalability, making it ideal for industrial production of devices such as FinFETs, nanowire FETs, and multi-gate transistors.

10.6.1.1 Key steps in top-down fabrication

Substrate preparation is the first step in top-down fabrication, where a high-purity silicon wafer is selected as the substrate for building nano-FETs. In some cases, silicon-on-insulator (SOI) substrates are used to enhance device performance by reducing parasitic capacitance.

Thin film deposition is then carried out to deposit oxide, nitride, or metal layers onto the substrate. These films serve as active layers, masks, or insulating layers depending on the device design. Techniques such as chemical vapor deposition (CVD), physical vapor deposition (PVD), and ALD are used to ensure precise thickness control and uniform material distribution.

Lithography is a critical step in defining nano-scale patterns on the wafer. Photolithography uses ultraviolet (UV) or extreme ultraviolet (EUV) light to transfer patterns onto a photoresist-coated wafer, with EUV lithography enabling sub-10 nm resolution using 13.5 nm light (Chen et al., 2021). Electron beam lithography (EBL), an alternative technique, directly writes nano-scale patterns onto the wafer using a focused electron beam, achieving resolutions below 5 nm. However, EBL is slow and mainly used for research and low-volume production (Tersoff, 2021). The comparison of all lithography techniques for nano-FET fabrication is shown in Figure 10.12.

Etching follows lithography, where material is selectively removed to form nano-scale transistor structures. Reactive ion etching (RIE) is commonly used for anisotropic etching, which provides high aspect-ratio patterns. Atomic layer etching (ALE), on the other hand, enables atomic-level precision, making it essential for finely tuned nano-scale architectures (Lee et al., 2025).

Doping introduces controlled impurities into the silicon to modulate its electrical properties. The most common method is ion implantation, where dopant ions are accelerated and embedded into the substrate. A subsequent annealing step ensures that the dopants are activated, influencing the carrier concentration and threshold voltage of the FET (Novoselov et al., 2004).

Planarization is the final step, ensuring that the wafer surface is flat and uniform. Chemical mechanical planarization (CMP) is widely used to achieve a smooth surface, which is crucial for multi-layered structures in advanced nano-FETs (Kuhn, 2023).

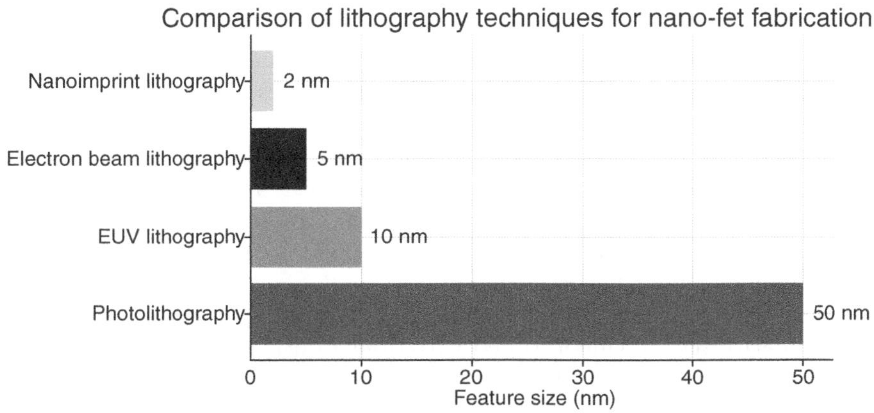

Figure 10.12 Comparison of all lithography techniques for nano-FET fabrication.

10.6.1.2 Applications of top-down techniques

FinFET fabrication is a key application of top-down techniques. Fin structures are formed using lithography and RIE. The etched silicon fins are then doped and surrounded by a gate material, providing better electrostatic control and reducing short-channel effects (Hisamoto et al., 2000).

Nanowire FETs are also fabricated using top-down approaches, where nano-scale wires are selectively etched from bulk silicon or formed through self-aligned etching techniques. These nanowires enable the development of Gate-All-Around (GAA FET architectures, which offer enhanced electrostatic control compared to FinFETs (Jariwala et al., 2014).

10.6.1.3 Advantages of top-down approaches

Scalability is one of the primary advantages of top-down fabrication, as these techniques are well-established for the mass production of nano-FETs. They ensure consistent performance and high yield, making them ideal for CMOS manufacturing.

Precision is another benefit, as top-down lithography and etching techniques enable feature sizes as small as a few nanometers, which is crucial for high-performance transistors. Compatibility with existing semiconductor technology makes top-down fabrication an attractive choice. As these techniques align with current CMOS manufacturing processes, they facilitate the seamless integration of new transistor architectures.

10.6.1.4 Challenges of top-down techniques

The high cost of advanced fabrication equipment, especially EUV lithography tools, poses a major challenge. The infrastructure required for sub-10 nm patterning involves significant financial investment, which can limit accessibility for smaller manufacturers.

Physical and technical constraints also hinder further feature size reduction. While lithography techniques continue to evolve, the approach is approaching fundamental diffraction limits, requiring new innovative solutions such as directed self-assembly (DSA) to sustain scaling trends (Akinwande et al., 2019).

Defects introduced during etching and doping can significantly affect device performance. Uncontrolled variations in dopant concentration, etch profiles, or planarization inconsistencies can lead to electrical variability, which impacts transistor reliability and yield (Novoselov et al., 2004).

10.6.2 Bottom-up approaches in nano-FET fabrication

Bottom-up approaches focus on assembling materials and structures atom by atom or molecule by molecule, enabling the creation of highly precise nano-scale architectures.

These methods harness natural forces, such as chemical reactions and self-assembly, to produce structures with desired electronic properties. Bottom-up fabrication is particularly valuable for creating novel materials like CNTs, graphene, and transition metal dichalcogenides (TMDs), which are integral to next-generation nano-FETs.

10.6.2.1 Key steps and methods in bottom-up fabrication

CVD is a widely used technique for growing thin films or nanostructures, such as CNTs and graphene. In this process, precursors in gaseous form react on a heated substrate, depositing material layer by layer. This method is essential for growing high-quality graphene for 2D FETs and for producing CNTs with uniform alignment and chirality control, which are crucial for CNTFET applications. A common example of this process is the decomposition of hydrocarbon gases, such as methane, at high temperatures, which results in the formation of a monolayer of graphene on a copper substrate.

Molecular beam epitaxy (MBE) is another PVD technique, in which atomic beams are directed onto a substrate in an ultra-high vacuum to enable atomic-level control. This method is widely used in fabricating quantum wells and heterostructures, as well as in building TMD-based FETs with precise layer stacking, leading to improved performance in nano-scale transistors (Tersoff, 2021).

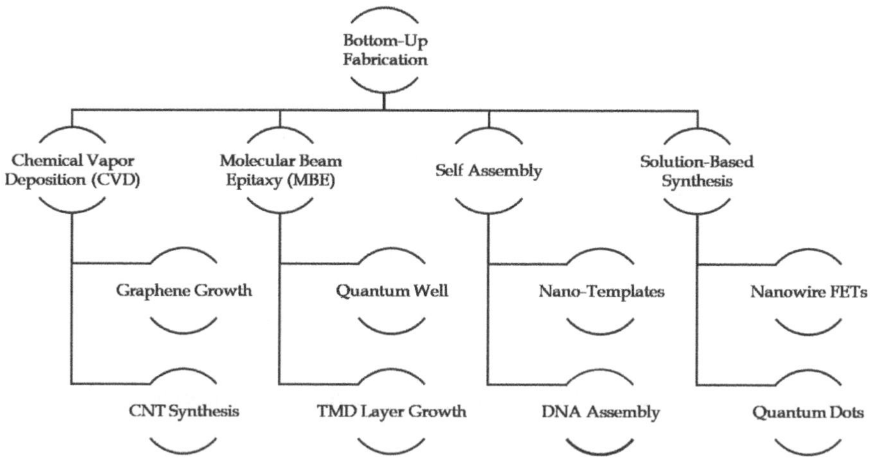

Figure 10.13 Bottom-up fabrication methods for nano-FETs.

Self-assembly techniques rely on molecular interactions to form ordered structures naturally. For example, block copolymer lithography is used to create templates for nanostructure growth, while DNA-directed assembly allows for the precise placement of CNTs and nanoparticles (Robertson, 2020). These techniques are particularly useful in creating periodic arrays of nanostructures, which are integrated into FET channels for improved functionality.

Solution-based synthesis uses chemical processes like sol-gel techniques to synthesize nanomaterials such as nanowires and quantum dots. This method is extensively used in producing semiconducting nanowires, which are then integrated into nanowire FETs for flexible or transparent electronic applications.

10.6.2.2 Advantages of bottom-up approaches

Bottom-up fabrication techniques enable the creation of defect-free materials with tailored properties, allowing for precise material engineering at the atomic level. In addition, they facilitate the integration of advanced materials, such as CNTs, graphene, and TMDs, which provide superior electrical, mechanical, and optical properties compared to traditional silicon. Another key advantage of bottom-up approaches is that methods like CVD are scalable for industrial applications, making them suitable for mass production in semiconductor manufacturing.

10.6.2.3 Challenges in bottom-up approaches

Despite their advantages, bottom-up approaches face several challenges in practical implementation. One of the primary issues is control over material properties. For instance, achieving uniform chirality in CNTs and precisely controlling the thickness of TMD layers remains difficult, impacting device performance. Another significant challenge is integration, as combining bottom-up fabricated materials with top-down processes is still a major hurdle due to differences in material compatibility and processing conditions. Finally, scalability remains an issue, as some techniques, like MBE, are time-consuming and not yet viable for large-scale production, making them less attractive for commercial semiconductor manufacturing.

10.6.2.4 Applications of bottom-up approaches in nano-fets

CNTFETs leverage bottom-up methods like CVD to grow high-purity CNTs, which are then aligned and integrated into FETs. These transistors offer high-speed switching and low-power operation, making them suitable for next-generation computing applications (Akinwande et al., 2019).

Graphene-based FETs benefit from CVD and exfoliation methods, which allow for the growth of high-quality graphene layers. These transistors exhibit exceptional carrier mobility, making them ideal for RF transistors and sensor applications (Novoselov et al., 2004).

TMD-based FETs are fabricated using MBE or CVD, enabling the growth of layered semiconductors like MoS_2. These materials enhance scalability and performance in high-performance FETs, making them a promising alternative to silicon-based transistors in future semiconductor technology (Jariwala et al., 2014).

10.6.3 Challenges in fabricating nano-FETs

Fabricating nano-FETs is a complex process that encounters numerous challenges due to the scale, materials, and technologies involved. These challenges stem from the need for atomic-level precision, material integrity, and reliable integration into industrial processes. Below is a detailed explanation of the key challenges, categorized for graduate and postgraduate students.

10.6.3.1 Scaling challenges and device dimensions

10.6.3.1.1 Extreme miniaturization

As transistor dimensions shrink below 5 nm, maintaining precise control over the physical and electrical properties of devices becomes increasingly difficult. Issues like variability in channel dimensions and gate lengths can lead to significant performance deviations.

10.6.3.1.2 Quantum effects

At nano-scales, quantum phenomena such as tunneling and electron wave interference dominate, causing leakage currents, increased power dissipation, and reduced gate control. These effects necessitate advanced materials and architectures like FinFETs or GAA transistors (Chau et al., 2005).

10.6.3.2 Material challenges

10.6.3.2.1 Defects in nanomaterials

Nanomaterials like graphene, CNTs, and TMDs often contain defects that degrade device performance. For example, variations in chirality in CNTs lead to inconsistent electrical properties.

10.6.3.2.2 Interface engineering

Poor interfaces between different materials, such as a high-k dielectric layer and the semiconductor channel, can introduce traps and degrade device mobility.

10.6.3.2.3 Integration of new materials

The integration of advanced materials like 2D semiconductors into CMOS-compatible processes remains difficult due to differences in deposition techniques and thermal budgets.

10.6.3.3 Process-related challenges

10.6.3.3.1 Lithographic limitations

Conventional lithography methods face resolution limits. Although EUV lithography enables sub-10 nm patterning, it is expensive and technically demanding, requiring precise alignment and high-quality masks.

10.6.3.3.2 Etching challenges

Maintaining anisotropy in etching while avoiding material damage at the nanoscale is complex. Over-etching or rough etch profiles can compromise device functionality.

10.6.3.3.3 Planarization

With multiple layers of materials in advanced architectures, achieving a perfectly flat surface for subsequent processing is increasingly challenging. CMP introduces defects if not precisely controlled.

10.6.3.4 Reliability and variability

Nano-FETs are highly sensitive to variations in doping concentrations, material properties, and fabrication processes, leading to device-to-device variability.

Phenomena like hot-carrier degradation, time-dependent dielectric breakdown, and electromigration become pronounced at nano-scales, impacting the lifespan and reliability of devices.

10.6.3.5 Manufacturing scalability and cost

Advanced fabrication tools, such as EUV lithography and ALD, are expensive and require significant investment in infrastructure.

Defects during fabrication can result in low yields, increasing the overall cost of manufacturing. As devices move toward 3D architectures (e.g., GAA FETs), assembling and interconnecting billions of transistors with minimal error becomes a daunting task (Doyle et al., 2003).

10.6.3.6 Thermal management

10.6.3.6.1 Heat dissipation

At smaller scales, power density increases, leading to significant challenges in dissipating heat. Poor thermal management can result in overheating, reducing device efficiency and reliability (Pop, 2010).

Figure 10.14 illustrates heat generation and dissipation in a densely packed nano-FET array, highlighting the thermal challenges in modern semiconductor devices. The structure consists of semiconductor transistors on a substrate, with the source, gate, and drain forming the core components of the transistor. Heat is primarily generated in the active region beneath the gate due to high electric fields and carrier scattering, as indicated by the red shaded area. The red arrows depict the flow of heat, which moves downward through the semiconductor material into the substrate, acting as a heat sink. Efficient heat dissipation is essential to prevent performance degradation, increased leakage currents, and reduced reliability in nano-FETs. However, as transistors shrink and packing density increases, thermal management becomes more challenging due to limited thermal conductivity in ultra-thin materials. Advanced engineering techniques, such as

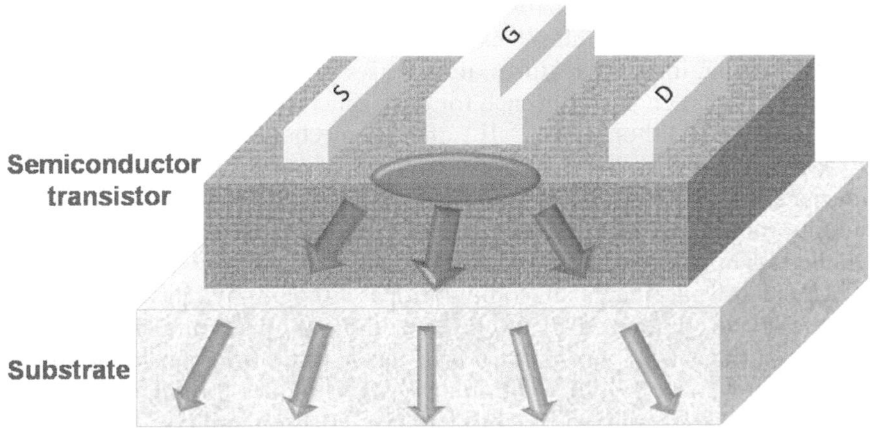

Figure 10.14 Visualization of heat generation and dissipation in a densely packed nano-FET array.

high-thermal-conductivity substrates and heat-spreading materials, are crucial for maintaining device stability and efficiency.

10.6.4 Advances in lithography for nanoelectronics

Advances in lithography have been pivotal in driving nanoelectronics forward, enabling the production of devices at ever-decreasing scales while maintaining performance and functionality. Modern lithography techniques have surpassed the traditional optical resolution limits defined by Rayleigh's criterion, utilizing novel approaches like EUV lithography, DSA, and multipatterning strategies.

EUV lithography operates at a wavelength of 13.5 nm, enabling sub-7 nm feature patterning. Unlike traditional deep ultraviolet (DUV) systems, EUV uses reflective optics to handle the short wavelength, but its adoption has been challenging due to complexities in source power, mask design, and photon-induced defects. Despite these hurdles, EUV is now a cornerstone of advanced semiconductor manufacturing, producing chips for applications such as artificial intelligence and 5G (Wurm et. al. 2014).

DSA leverages block copolymers that spontaneously arrange into highly ordered nanostructures under thermal or chemical treatment. This technique supplements conventional lithography, offering cost-effective and scalable solutions for creating periodic nanostructures with features below 5 nm. DSA is particularly promising for patterning regular structures like memory arrays and interconnects (Wu et al., 2023).

Multipatterning techniques, such as double and quadruple patterning, extend the resolution capabilities of conventional lithography tools by sequentially applying multiple exposure and etching cycles. While effective for fabricating intricate features, it increases process complexity and costs, making it a challenging approach for next-generation semiconductor nodes.

Nanoimprint lithography (NIL) offers a mechanical stamping approach to patterning, bypassing the resolution limits of optical methods. It is highly efficient for applications in photonic devices and biosensors, though its adoption in mainstream semiconductor fabrication has been limited due to challenges in defect control and throughput.

Recent advancements in computational lithography, including optical proximity correction (OPC) and inverse lithography techniques (ILTs), have enhanced the accuracy of pattern transfer by optimizing mask designs and illumination strategies. In addition, hybrid techniques combining EUV with NIL or DSA are being explored to balance precision and cost-effectiveness. These breakthroughs collectively enable the fabrication of transistors, interconnects, and memory devices at dimensions previously unattainable, driving the innovation of nanoelectronics (Shin et al., 2021).

10.7 APPLICATIONS OF NANO-FETS

Nanotechnology has revolutionized the field of electronics, with nano-scale field-effect transistors (nano-FETs) emerging as fundamental components in various cutting-edge applications. Due to their superior electrical performance, reduced power consumption, and enhanced sensitivity, nano-FETs play a crucial role in modern computing, energy-efficient electronics, sensing technologies, and emerging fields such as neuromorphic computing and quantum devices. This chapter explores the key applications of nano-FETs, highlighting their role in high-speed computing, low-power electronics, sensor technology, and emerging fields.

10.7.1 Nano-FETs in high-speed computing

High-speed computing demands transistors with superior switching capabilities, minimal delay, and low power dissipation. Nano-FETs, particularly those based on materials such as CNTs, graphene, and 2D TMDs, offer remarkable advantages over conventional silicon-based FETs. These transistors exhibit ultra-fast carrier mobility, high current density, and reduced short-channel effects, making them ideal for high-performance computing applications.

For instance, CNTFETs have demonstrated exceptional electron mobility (~100,000 cm^2/Vs) and ballistic transport properties, reducing power dissipation and enhancing processing speeds (Schwierz, 2016). Similarly, graphene-based FETs (GFETs) leverage the near-ballistic transport of charge carriers, enabling terahertz-speed transistors suitable for next-generation processors (Lemme et al., 2022). Recent research also highlights *2D-material-based FETs*, such as MoS_2 and tungsten diselenide (WSe_2), for their high on/off current ratio and superior electrostatic control, addressing challenges in miniaturization (Das et al., 2019).

10.7.2 Role in low-power electronics

With the growing demand for energy-efficient devices, nano-FETs have become critical in enabling low-power electronic systems. By leveraging unique nanoscale properties, these transistors achieve ultra-low subthreshold swing, reduced leakage currents, and improved energy efficiency, making them essential for portable and battery-operated devices.

One significant advancement is the development of TFETs, which utilize band-to-band tunneling for carrier transport, achieving subthreshold swings below 60 mV/decade—significantly lower than conventional MOSFETs (Seabaugh & Zhang, 2010). In addition, negative capacitance FETs (NC-FETs), based on ferroelectric materials, have demonstrated improved power efficiency by amplifying gate control, reducing supply voltage requirements (Salahuddin & Datta, 2008).

10.7.3 Sensors and biosensors using nano-FETs

Nano-FETs have revolutionized the field of sensing, offering highly sensitive detection mechanisms for chemical, biological, and environmental applications. Their small size, high surface-to-volume ratio, and tunable electrical properties enable real-time, label-free sensing with exceptional sensitivity.

10.7.3.1 Chemical and gas sensors

Nano-FET-based gas sensors, particularly *semiconducting metal oxide FETs*, exhibit superior performance in detecting trace gas molecules, such as NO_2, CO, and H_2S, due to their fast response time and high selectivity (Maashi et al., 2024). Graphene and CNT-based FET sensors further enhance sensitivity through enhanced adsorption of gas molecules and charge transfer mechanisms.

10.7.3.2 Biosensors

Nano-FETs are widely used for biomolecular detection, including DNA, proteins, and viruses. For instance, silicon nanowire FETs (SiNW-FETs) have demonstrated attomolar-level sensitivity in detecting cancer biomarkers, enabling early-stage disease diagnosis (Stern et al., 2007). Furthermore,

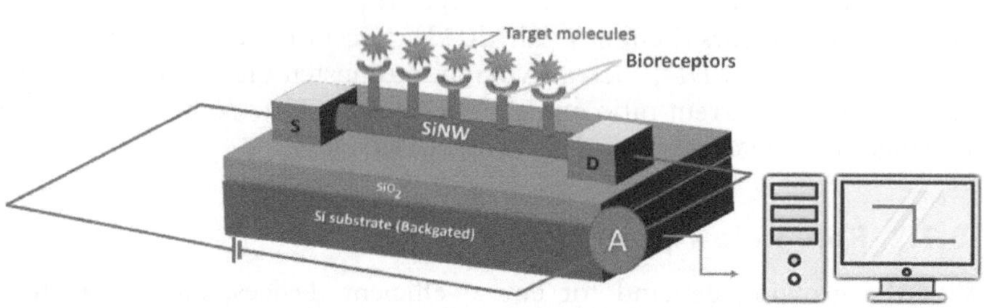

Figure 10.15 A graphical depiction of nano-FET-based biosensors illustrates the interaction between target biomolecules and nanoscale transistors, highlighting charge-induced conductivity changes.

2D-material-based biosensors, such as MoS_2-FETs, have been developed for real-time glucose monitoring and virus detection (Singhal et al., 2018).

Figure 10.15 represents a nano-FET-based biosensor that detects target biomolecules through changes in conductivity induced by surface interactions. At the core of the device is a silicon nanowire, which acts as the active sensing channel between the source (S) and drain (D) terminals. The nanowire is functionalized with bioreceptors that specifically bind to target molecules, triggering a localized charge effect. This interaction alters the electrical properties of the nanowire, leading to a measurable change in current or conductance. The buried SiO_2 layer serves as an insulating layer, while the Si substrate with back-gating allows for additional tuning of the device's sensitivity. The detected signal is amplified (A) and transmitted to a computer for processing and analysis. Such biosensors are highly sensitive and capable of real-time detection, making them valuable for applications in medical diagnostics, environmental monitoring, and biochemical research.

10.7.4 Emerging applications

As Nano-FET technology continues to evolve, emerging applications span beyond traditional electronics and sensing, venturing into neuromorphic computing, quantum information processing, and flexible electronics.

10.7.4.1 Neuromorphic computing

Nano-FETs have shown promise in mimicking synaptic behavior for brain-inspired computing. *Memristive FETs*, integrating resistive switching materials, have been used to develop artificial synapses with energy-efficient learning capabilities (Prezioso et al., 2015).

10.7.4.2 *Quantum and spintronics applications*

Quantum-dot-based nano-FETs and *spin-FETs* are paving the way for next-generation quantum computing and spintronic devices. These transistors exploit spin-polarized currents and quantum tunneling effects to enable faster, more secure computing architectures.

10.7.4.3 *Flexible and wearable electronics*

Nano-FETs fabricated on flexible substrates, such as *organic semiconductors and 2D materials*, have enabled the development of lightweight, stretchable electronic devices for wearable health monitoring and electronic skin.

Nano-FETs have emerged as transformative components across multiple domains, from high-speed computing and low-power electronics to advanced sensors and futuristic applications. Their continued development promises groundbreaking advancements in computing efficiency, healthcare, and wearable technology.

10.8 FUTURE TRENDS AND CHALLENGES

As FETs continue to scale down in size and evolve in functionality, researchers face both significant opportunities and challenges. The demand for higher processing speeds, lower power consumption, and enhanced device functionality is driving the exploration of new frontiers in transistor technology. In this chapter, we discuss the potential impact of quantum computing on FET development, the ongoing efforts to overcome material limitations, and the latest innovations in transistor design that could shape the future of nanoelectronics.

10.8.1 Opportunities in quantum computing

Quantum computing has the potential to revolutionize computation by leveraging the principles of superposition and entanglement, offering exponential speedups over classical computing for certain problems. Traditional transistor-based computing is approaching its physical limits, making quantum computing an attractive alternative for future high-performance applications. Quantum field-effect transistors (QFETs) and spintronic FETs are emerging as promising candidates for integrating quantum mechanics into transistor technology. These devices utilize quantum dots, superconducting materials, and topological insulators to manipulate quantum states and improve computational efficiency. However, significant challenges remain, including maintaining quantum coherence, reducing error rates, and developing scalable fabrication techniques. If these challenges can be addressed, quantum-enhanced transistors could play a key role in the next generation of computing architectures (Lund et al., 2023).

10.8.2 Overcoming material limitations

As transistor sizes shrink to the atomic scale, the limitations of traditional silicon-based materials become increasingly apparent. Issues such as electron scattering, increased leakage currents, and short-channel effects hinder the performance of nanoscale FETs. To address these challenges, researchers are exploring alternative materials, including 2D materials like graphene and TMDs, as well as high-mobility materials such as germanium and III–V semiconductors. In addition, the integration of high-k dielectrics, ferroelectric materials, and novel heterostructures has shown promise in improving transistor performance and reliability. Advanced fabrication techniques, including ALD and MBE, are also being optimized to enable precise control over material properties at the nanoscale. Despite these advancements, challenges remain in large-scale manufacturability, interface engineering, and long-term stability of these emerging materials in commercial applications (Radisavljevic & Kis, 2023).

10.8.3 Innovations in FET design

In response to the limitations of traditional transistor scaling, researchers are actively exploring novel FET architectures that offer enhanced performance, efficiency, and adaptability to new computing paradigms. One of the most promising innovations is the development of GAA transistors, which provide superior electrostatic control compared to conventional FinFETs. Another emerging design is the NC-FET, which utilizes ferroelectric materials to reduce subthreshold swing, thereby improving power efficiency. Furthermore, flexible and stretchable transistors are being developed for wearable electronics and biomedical applications, enabling new functionalities in human-machine interfaces. These advancements in FET design are expected to push the boundaries of transistor technology, paving the way for ultra-low-power computing, artificial intelligence acceleration, and next-generation communication systems (Kim et al., 2024).

The future of FET technology is poised at the intersection of quantum computing, material innovation, and architectural breakthroughs. While significant challenges remain, the rapid pace of research and development in these areas offers exciting opportunities for the continued evolution of nanoelectronics. By addressing the limitations of current materials, refining transistor architectures, and exploring quantum-enhanced computing, researchers can unlock new possibilities for high-performance, energy-efficient, and scalable electronic devices.

REFERENCES

Akinwande, D., Cric, H., Wang, C., Serna, M. I., Goossens, S., Li, L., Wong, H-S P., Koppens, F.H L. (2019). Graphene and two-dimensional materials for silicon technology. *Nature*, 573(7775), 507–518.

Al-Turk, S. (2022). Electronic Transport in Mixed-dimensional van der Waals Heterostructures. University of Toronto.

Chau, R., Datta, S., Doczy, M., Doyle, B., Kavalieros, J., & Metz, M. (2005). Benchmarking nanotechnology for high-performance and low-power logic transistor applications. *IEEE Transactions on Nanotechnology*, 4(2), 153–158.

Chaure, N. B., Basova, T., Ray, A. K., Gürek. A. G., & Ahsen, V. (2009). Memory effects in thin film organic transistor characteristics. *Journal of Physics D: Applied Physics*, 42, 12.

Chaure, N. B., Basova, T., Zahedi, M., Ray, A. K., Sharma, A. K., Durmuş, M., & Ahsen, V. (2010a). Solution processed tetrasubstituted zinc phthalocyanine as an active layer in organic field effect transistors. *Journal of Applied Physics*,107, 114503.

Chaure, N. B., Cammidge, A. N., Chambrier, I., Cook, M. J., Cain, M. G., Murphy, C. E., Pal, C., & Ray, A. K. (2011). High-mobility solution-processed copper phthalocyanine-based organic field-effect transistors. *Science and Technology of Advanced Materials*, 12, 025001–025007.

Chaure, N. B., Cammidge, A. N., Chambrier, I., Cook, M. J., & Ray, A. K. (2015). A tetrabenzotriazaporphyrin based organic thin film transistor: comparison with a device of the phthalocyanine analogue. *ECS Journal of Solid State Science and Technology*, 4 (4), P3086–P3090.

Chaure, N. B., Pal, C., Barard, S., Kreouzis, T., Ray, A. K., Cammidge, A. N., Chambrier, I., Cook, M. J., Murphy, C. E., & Cain, M. G. (2012). A liquid crystalline copper phthalocyanine derivative for high performance organic thin film transistors. *Journal of Materials Chemistry*, 22, 19179.

Chaure, N. B., Sosa-Sanchez, J. L., Cammidge, A. N., Cook, M. J., & A. K. Ray, A. K. (2010b). Solution processable lutetium phthalocyanine organic field-effect transistors. *Organic Electronics*, 11, 434–438.

Chen, Y., Shu, Z., Zhang, S., Zeng P., Liang, H., Zheng M., and Duan, H., (2021) Sub-10 nm fabrication: methods and applications. *International Journal of Extreme Manufacturing*, 3, 032002.

Das, S., Sebastian, A., Pop, E., McClellan, C. J., Franklin, A. D., Grasser, T., Knobloch, T., Illarionov, Y., Penumatcha, A. V., Appenzeller, J., Chen, Z., Zhu, W., Asselberghs, I., Li, L., Avci, U. E., Bhat, N., Anthopoulos T. D., and Singh, R., (2021). Transistors based on two-dimensional materials for future integrated circuits. *Nature Electronics, 4*, 786–799.

Dash, T., & Jena, J. (2024). *Strain Engineering for Highly Scaled MOSFETs*. Wiley Online Library.

Doyle, B. S., Datta, S., & Doczy, M. (2003). High performance fully-depleted tri-gate CMOS transistors. *IEEE Electron Device Letters*, 24(4), 263–265.

Faris, T., Basova, T., Chaure, N. B., Sharma, A. K., Durmus, M., Ahsen, V., & Ray, A. K. (2014). Effects of annealing on device parameters of organic field effect transistors using liquidcrystalline tetrasubstituted zinc phthalocyanine. *EPL*, 106, 58002.

Fei, A. (2024). *A Review of the Development of Transistor Technology and Its Application in Electronic Devices*. Semantic Scholar.

Hisamoto, D., Lee, W-C., Kedzierski, J., Takeuchi, H., Asano, K., Kuo, C., Anderson, E., King, T-J., Bokor, J., and Hu, C., (2000). FinFET—a self-aligned double-gate MOSFET. *IEEE Transactions on Electron Devices*, 47, 12.

Jariwala, D., Sangwan, V. K., Lincoln, J. Lauhon, Tobin J. Marks, Hersam, M. C., (2014). Emerging device applications for semiconducting two-dimensional transition metal Dichalcogenides, *ACS NANO*, 2, 1102–1120.

Jiang, H., Zhang, X., Chen, K., He, X., Liu, Y., Yu, H., & Gao, L. (2025). *Two-dimensional Czochralski Growth of Single-crystal MoS2*. Nature Materials.

Kalra, S., Kannan, N., & Kumar, M. J. (2024). *A Partially Gated Regenerative Feedback Device for Ultra-high Sensitivity Biosensing Applications*. IEEE Transactions on Nanotechnology.

Kim, H., Park, J., & Lee, S. (2024). Next-generation FET Architectures: Gate-all-around and Beyond. IEEE Transactions on Nanotechnology.

Kuhn, K. (2023). Scaling Beyond CMOS: Challenges and Opportunities. IEEE Microelectronics Journal.

Lee, T-Y., Chen, P-T., Huang, C-C., Chen, H.-C., Chen, L-Y., Lee, P-T., Chen, F-C., Horng R.-H., and Kuo, H-C., (2025). Advances in core technologies for semiconductor manufacturing: applications and challenges of atomic layer etching, neutral beam etching and atomic layer deposition. *Nanoscale Advances,* 7, 2796–2817.

Lemme, M. C., Akinwande, D., Huyghebaert, D., & Stampfer, C., (2022). 2D materials for future heterogeneous electronics. *Nature Communication,* 13, 1392.

Lenka, T. R., & Nguyen, H. P. T. (2024). *Nanoelectronic Devices and Applications*. Google Books.

Londhe, P., Chaure, N. B., & Athawale, A. (2021). Interface engineering of gate dielectrics with multifunctional self-assembled monolayers in copper phthalocyanine based organic field-effect transistors. *Materials Science and Engineering: B*, 273, 115397.

Lund, A., White, A. G., & Ralph, T. C. (2023). Advancements in Quantum Transistor Technology: Overcoming Coherence Challenges. Nature Physics.

Ma, L., Wang, Y., & Liu, Y. (2024). van der Waals Contact for Two-Dimensional Transition Metal Dichalcogenides. Chemical Reviews, ACS Publications.

Maashi F., Aljarb, A., and Al-Jawhari, H., (2024). Flexible and disposable gas sensors based on two-dimensional materials. *Key Engineering Materials*, 977, 85–91.

Mo, F., Scognamiglio, L., & Spano, C. E. (2025). *Simulation Study on the Impact of Miniaturization in 3 nm Node 3D Junctionless Transistors*. IEEE Transactions on Nanotechnology.

Mohanbabu, A., & Maheswari, S. (2024). *Advancements in GaN Technologies: Power, RF, Digital and Quantum Applications*. Bentham Direct.

Novoselov, K. S.,, A. K. Geim, S. V. Morozov, D. Jiang, Y. Zhang, S. V. Dubonos, I. V. Grigorieva, and A. A. Firsov, (2004). Electric field effect in atomically thin carbon layers. *Science*. 306, 666–669.

Pop, E. (2010). Energy dissipation and transport in nanoscale devices. *Nano Research*, 3 (3), 147–169.

Prezioso, M., Merrikh-Bayat, F.,Hoskins, B. D., Adam, G. C., Likharev K. K., & Strukov, D. B., (2015). Training and operation of an integrated neuromorphic network based on memristors. *Nature*, 521(7550), 61–64.

Radhakrishnan, S., Vishnu, S. N. S., Ahmed, S. I., Thiruvengadathan, R., (2023). Effects of Channel Length Scaling on the Electrical Characteristics of Multilayer MoS2 Field Effect Transistor. *Micromachines*, MDPI., 14, 275.

Radisavljevic, B., & Kis, A. (2023). 2D materials in transistor design: Challenges and opportunities. *ACS Nano*, 17(4), 5632–5648.

Robertson, J. (2020). High-k Dielectrics for Advanced Nanoelectronics. *Materials Today*.

Saini, S., & Tyagi, H. K. (2024). *Performance of Si and GaAs GAA Nanowire FET for 8 nm Channel Length Using 3D Full Quantum Transport Simulation*. Springer.

Salahuddin, S., & Datta, S. (2008). Use of negative capacitance to provide ultra-low power transistors. *Nano Letters*, 8(2), 405–410.

Schwierz, F. (2016). Graphene transistors: Status, prospects, and problems. *Proceedings of the IEEE*, 104(7), 1491–1516.

Seabaugh, A. C., & Zhang, Q. (2010). Low-voltage tunnel transistors for beyond CMOS logic. *Proceedings of the IEEE*, 98(12), 2095–2110.

Shah, A., & Kashyap, A. (2024). *Nanoribbon of Zigzag Silicene for Spin Field Effect Transistor with Low Voltage Bias*. IEEE International Magnetic Conference.

Shin, Y. (2021). Computational lithography using machine learning models. *IPSJ Transactions on System LSI Design Methodology*, 14, 2–10.

Singhal, C., Khanuja, M., Chaudhary, N., Pundir, C., & Narang, J. (2018). Detection of chikungunya virus DNA using two-dimensional MoS2 nanosheets based disposable biosensor. *Scientific Reports, 8*(1), 1–11.

Stern, E., Vacic, A., Rajan, N. K., Criscione, J. M., Park, J., Ilic, B. R., Mooney, D. J., Reed M. A., and Fahmy, T. M., (2009). Label-free biomarker detection from whole blood. *Nature Nanotechnology*, 5, 138–146.

Tamersit, K., Madan, J., Kouzou, A., Pandey, R., & Kennel, R. (2022). *Role of Junctionless Mode in Improving the Photosensitivity of Sub-10 nm Carbon Nanotube/Nanoribbon Field-effect Phototransistors: Quantum Simulation, Performance Analysis, and Insights*. Nanomaterials, MDPI.

Tersoff, J. (2021). Carbon Nanotubes for Nanoelectronics. Nature Nanotechnology.

Wu, M., Wang, D., Wan, L-J. (2016). Directed block copolymer self-assembly implemented via surface-embedded electrets, *Nature communication*, 7, 10752.

Wurm, S. (2014). EUV lithography – Progress, challenges and outlook. *Proceedings of SPIE*, 9231, 923103-2.

Xu, H., Gan, W., Guo, S., & Zhang, S. (2024). *Machine Learning-based Compact Modeling of Silicon Cold Source Field-effect Transistors*. IEEE Transactions on Nanotechnology.

Yang, J., Chen, C., Chen, X., Wu, Z., & Yang, S. (2024). *Ballistic Transport in Planar 2-D ZnXNTEXPRESERVE0 (X Si, Ge) Transistors with Ultrahigh On-State Current*. IEEE Transactions on Electron Devices.

Zhang, H., Chi, M., Tian, S., Liang, T., Liu, J., Zhang, X., & Wan, L. (2024). Giant Piezotronic Effect in Ferroelectric Field-effect Transistor. Nano Research.

Index

For Product Safety Concerns and Information please contact our
EU representative GPSR@taylorandfrancis.com Taylor & Francis
Verlag GmbH, Kaufingerstraße 24, 80331 München, Germany